Signals and Communication Technology

This series is devoted to fundamentals and applications of modern methods of signal processing and cutting-edge communication technologies. The main topics are information and signal theory, acoustical signal processing, image processing and multimedia systems, mobile and wireless communications, and computer and communication networks. Volumes in the series address researchers in academia and industrial R&D departments. The series is application-oriented. The level of presentation of each individual volume, however, depends on the subject and can range from practical to scientific.

Indexing: All books in "Signals and Communication Technology" are indexed by Scopus and zbMATH

For general information about this book series, comments or suggestions, please contact Mary James at mary.james@springer.com or Ramesh Nath Premnath at ramesh.premnath@springer.com.

Yasir Ahmed

Recipes for Communication and Signal Processing

Springer

Yasir Ahmed
Independent 5G/Solar Energy Consultant
Founder, RAYmaps
Karachi, Pakistan

ISSN 1860-4862 ISSN 1860-4870 (electronic)
Signals and Communication Technology
ISBN 978-981-99-2919-1 ISBN 978-981-99-2917-7 (eBook)
https://doi.org/10.1007/978-981-99-2917-7

This Springer imprint is published by the registered company Springer Nature Singapore Pte Ltd.
The registered company address is: 152 Beach Road, #21-01/04 Gateway East, Singapore 189721, Singapore

کوئی قابل ہو تو ہم شانِ کئی دیتے ہیں
ڈھونڈنے والوں کو دنیا بھی نئی دیتے ہیں

Allama Iqbal

Nobody ever figures out what life is all about, and it doesn't matter. Explore the world. Nearly everything is really interesting if you go into it deeply enough.
Study hard what interests you the most in the most undisciplined, irreverent, and original manner possible.
— Richard P. Feynman

Dedicated to my dear daughter Zahra, and to Ammi and Abu for always standing by me.

Foreword

The field of wireless communication has enormously evolved in the last three decades. However, this book explains the elementary principles of wireless communications in simple words for a mass audience with little or no prior knowledge. At the same time, this book also intends to target students, teachers, researchers and professionals. The basic concepts are explained with simple language so that they are easy to grasp and understand. Illustrative examples are included for explaining the concepts. This book provides MATLAB and Python codes for beginners, which help in understanding and visualizing the working of different components of wireless systems. This book covers some novel topics such as full-duplex, reconfigurable intelligent surfaces, and index modulation.

The capacity of wireless channels has always been a topic of great interest for the researchers, and the discussion of capacity is incomplete without considering the Shannon Capacity. The author of the book explains the Shannon Capacity of different technologies, i.e., GSM, CDMA, LTE and 5G. In practical communication systems, often arrays of antennas are used instead of a single monopole or dipole antenna. Fundamentals of different types of antenna arrays such as linear, rectangular, and circular types are explained in this book with handy MATLAB code. This book also addresses the classical problem of finding the phase and frequency of a signal embedded in noise. In the last part of the book, the author has brought attention to some of the advanced topics of the modern era such as millimeter waves and concerns about health risks of 5G.

Muhammad Usman Sheikh
System Performance Specialist
Nokia
Espoo, Finland

Preface

My first experience at building a communication system was during the final year of my bachelor's program, where me and my project partner decided to build a Power Line Carrier System. After a lot of trial and error we were able to build a half-duplex system that worked over a distance of 100 m. We could not build a full-duplex system due to the limitations of the FSK chip we used. The main learning from this exercise was that a communication channel, a power line in this case, has both noise and interference and we have to somehow filter that out. Received power is distant dependent; the further away the receiver from the transmitter, the weaker the signal received. This work resulted in a paper in the All Pakistan IEEE Students Seminar, 1998.

As I moved into the industry after graduation I learnt that things were not that simple. There were tight deadlines and customer expectations to be met. At Data Communication and Controls, Karachi, I had the first experience of using microcontrollers to solve some real-world problems. I built two significant pieces of hardware; an eight channel multiplexer and demultiplexer that could send morse code over long distances and an infrared data logger that worked with a utility meter. The major learning was that on trial day "**anything that can go wrong will go wrong**." But if it works, and works for years, without any major bugs, it is sheer joy and amazement for the customer and you are bound to get repeat orders.

So, after getting my hands dirty and gaining some experience I decided to move to Virginia Tech for my MS in wireless communications. Soon after landing in Blacksburg VA I was hired by Mobile and Portable Radio Group as Research Assistant. My first major task was to study the problem of cochannel interference in GSM systems. I worked on this problem tirelessly and eventually showed that two MSK signals can be jointly detected provided there is a phase offset between them. This resulted in a publication in Globecomm 2003. But I was more interested in Space Time Block Codes and the kind of work that Alamouti, Jafarkhani and Tarokh were doing. I was one of the first ones to work on STBCs for eight transmit antennas and published my work in the Asilomar Conference, 2002.

After I had completed my MS degree program I got a cold call from Qualcomm that they had some openings in the ASIC Systems Test department and wanted to interview me on-site. So I flew to California the second time and got an offer a few weeks later. This was an exciting time as Qualcomm was migrating from CDMA-1x to WCDMA and expanding its footprint to Europe. My main responsibility was performance and conformance test of modems and later on I was also involved in field testing in Europe. The main learning for me here was that in a lab like MPRG it was probably ok to work on the physical layer only but once you go out into the field you have to look at the full stack.

After working in the industry for a while I realized my true calling was academia. So I moved back to Pakistan and started teaching and I even had a short stint in a Ph.D. program. During this time I worked on propagation modeling and measurement and published a conference paper on narrowband wireless channel modeling and another on WiMAX network planning using Google maps. There was also some work done on patch antenna design which resulted in journal publication in International Journal of Antennas and Propagation (IJAP). I also developed a ray-tracing engine in MATLAB using the Shooting and Bouncing Ray (SBR) method. This is still work in progress, and I intend to improve the accuracy and runtime using the latest techniques available in literature.

But I felt that all my research work was lying dormant deep in some library and was not accessible to all. Besides I felt that research papers are full of mathematical details and complex terminology which is intimidating to a newbie. So I started putting all the content that I had in the form of blog posts, simple and piecemeal. MATLAB, Octave and Python code was added to the blog posts to make them more interactive. Slowly but surely, it started to gain traction, and users from all over the world were visiting my blog, asking questions and also posting suggestions and improvements. So, the papers turned into a blog and the blog turned into a book. I present to you Recipes for Communication and Signal Processing, a name suggested by Dr. Jeff Reed in one of the several discussions we had. Thank you!

Karachi, Pakistan Yasir Ahmed

Acknowledgements

I would like to thank my teachers at Virginia Tech who instilled me with the passion for learning wireless communications. These include Dr. Jeffrey H. Reed, Dr. William H. Tranter, Dr. R. Michael Buehrer, Dr. Brian D. Woerner and Dr. Theodore S. Rappaport. It was in fact Dr. Rappaport's book that introduced me to wireless communications and inspired me to apply to the MS program at Virginia Tech. His lectures on cellular communications, particularly CDMA technology, were also very fascinating.

I would also like to mention my Stochastic Signals and Systems teacher Dr. Robert J. Boyle who encouraged me to apply to Mobile and Portable Radio Group (MPRG) where I later on worked as Research Assistant. Dr. Boyle noticed my interest in antenna arrays and as luck would have it there were plenty of research opportunities in antenna arrays in those days. MPRG continuously supported me and funded during the two years I was at Virginia Tech. MPRG was like a home away from home.

This was the time that the field of MIMO was buzzing with activity and S. M. Alamouti had just proposed his technique on transmit diversity. There was also some seminal work coming out from Bell Labs. This got me excited and I started working on Space Time Block Codes (STBC) which ultimately resulted in a full paper in 36th Asilomar Conference on Signals Systems and Computers. The collaboration with MPRG continued even after I graduated from Virginia Tech and has resulted in 12 research publications.

Last but not the least I would like to thank my advisor Dr. Jeff Reed for his continuous support and encouragement. When I started my blog about ten years back he was one of the first ones to support it and his comments have been very helpful over the years. He has a very deep understanding of wireless communications and a knack for presenting complex concepts in the most simplified form. Dr. Reed has a very strong bond with his students and I remember the times he used to invite us to his place for Thanksgiving dinners.

How can I forget the times I had on the cricket field, first at Virginia Tech and later on at University of California San Diego while I was working for Qualcomm Inc. Cricket was a great release for me from the pressures of university life and work life. I made great friends on and off the field including Ramiz Taqi, Srikanth Nathela, Prabhakar Thanikasalam, Aditya Gadre, Pushkar Ogale to name a few. Without them life in Virginia and California would not have been so much fun. I would also like to mention Sajjad, Hammad and Farooq who were a great support network in the USA and Dr. Usman Sheikh, at Nokia, Finland, for his valuable insights and suggestions.

Prologue: Ibn al-Haytham to Maxwell—A Long Road

As the Chinese proverb says "*The journey of a thousand miles begins with a single step.*" The journey that started with Ibn al-Haytham experimenting with his Camera Obscura in the eleventh century was completed eight hundred years later by James Clerk Maxwell and Heinrich Hertz. While Maxwell laid down the mathematical framework that described the behavior of Electromagnetic waves, Hertz conclusively proved the existing of these invisible waves through his experiments. There were several scientists on the way that played a crucial part in development of this electromagnetic theory such as Gauss, Faraday and Ampere. Then there were others such as Huygens, Fresnel and Young who worked on nature of light, which was not known to be an electromagnetic wave at that time. Once the theory of electromagnetic wave propagation was in place there was rapid progress in many fields, particularly in wireless communications (wireless telegraph, radio, radar, etc.).

Maxwell's equations that were proposed in 1861 were initially quite circuitous and were not well accepted. But later on, these equations were simplified into the form we now know by Oliver Heaviside. There are still two popular forms of the equations, the integral form and the differential form. We present the integral form of these equations in this article as it is more intuitive and is also easier to represent graphically. The differential form requires understanding of the concepts of divergence and curl and we skip in this article. The main take away from these equations (presented below) is that a changing electric field produces a magnetic field and a changing magnetic field produces an electric field and they are always perpendicular to each other and to the direction of propagation. Another important result is that magnetic monopoles do not exist (simply put a magnet, however small, always has a north and south pole).

	MAXWELL'S EQUATIONS IN INTEGRAL FORM	
LHS	$X = E$	$X = B$
$\oint X.dA$	Gauss's Law (1813) $\oint E.dA = \frac{Q_{inside}}{\varepsilon_o}$ The Electric field through a closed area is equal to the total charge inside of the area divided by ε_o.	Gauss's Law for Magnetism (1813) $\oint B.dA = 0$ The Magnetic field through a closed surface is zero (as many field lines going out as going in). It means that magnetic monopoles do not exist.
$\oint X.dl$	Faraday's Law (1831) $\oint E.dl = -\int \frac{\partial B}{\partial t}.dA$ The Electric field around a closed loop is just equal to the minus of the rate of change of Magnetic field through the loop.	Maxwell - Ampere's Law (1861) $\oint B.dl = \mu_o I + \mu_o \epsilon_o \int \frac{\partial E}{\partial t}.dA$ The Magnetic field around a closed loop is equal to rate of change of Electric field through the loop times $\mu_o \epsilon_o$ plus the Electric current in the loop times μ_o.

Maxwell's equations in integral form

Note

1. The dot product with a line segment means that only that component of the field vector is effective that is along the line segment. On the other hand, the dot product with a surface means that only that component is considered that is perpendicular to the surface (since the unit vector of a surface is perpendicular to the surface). It means that only those field components are considered that are going perpendicularly in or out of the surface.
2. For more on history of Maxwell equations visit IEEE Spectrum and for a detailed explanation of the various forms of the Maxwell's equations visit this page [https://spectrum.ieee.org/the-long-road-to-maxwells-equations].
3. In modern electromagnetic simulation software the differential form is preferred and the algorithm used is called finite difference time domain (FDTD). However, if the area of interest is quite large (with respect to the wavelength) then the FDTD method becomes prohibitively complex and another method known as ray-tracing is used. Please do check out the ray-tracing engine that we have developed. Ray-tracing is becoming increasingly important in RF Planning of Telecom Networks.

Contents

About the Author

Yasir Ahmed has more than 20 years of experience in various organizations in Pakistan, Europe, and the USA in both Engineering and Management roles. He worked as a Research Assistant in the Mobile and Portable Radio Group (MPRG) of Virginia Tech under the supervision of Dr. Jeff Reed and was one of the first researchers to propose Space Time Block Codes (STBCs) for eight transmit antennas. The collaboration with MPRG has continued over the years and has resulted in 12 research publications and a book on Wireless Communications.

Yasir worked as GM SEED at Ignite National Technology Fund, Pakistan, a company involved in supporting innovation and entrepreneurship eco-system in the country. He previously worked for Qualcomm USA, leading the physical layer performance and conformance testing of GSM/UMTS modems, and for COMSATS Islamabad as Assistant Professor, teaching various subjects in the Telecom and Networks area. He was part of the Ignite team that evaluated multi-billion-rupee NIC and DigiSkills programs and has also helped fund a number of startups that have gone on to become successful commercial ventures.

About the Author

Chapter 1
The Wireless Channel

1.1 Introduction

The most commonly used wireless channel model is Additive White Gaussian Noise (AWGN) model but it's quite a misleading notion as AWGN has nothing to do with the channel. It is in fact the noise generated by the receiver front end which is added to the received signal. The noise power depends upon the bandwidth of the receiver and the temperature. The noise is white, meaning that its power spectral density is flat, and it has a Gaussian distribution in the time domain. The ratio of signal power to noise power is commonly referred to as signal to noise ratio (SNR) and is an important metric that defines the performance of the receiver. Higher the SNR lower is the bit error rate and more bits can be passed through the channel.

Some of the other impairments that a wireless signal has to undergo are fading, interference, diffraction and scattering. Fading occurs when multiple copies of a signal, after reflection and refraction, arrive at the receiver and add constructively or destructively. A fade can be anywhere from 10 to 30 dB deep or even worse. On a linear scale it means the signal amplitude may be 10 times to 1000 times lower than no fading case. The two most common distributions used to describe temporal variations of a fading channel are Rayleigh and Ricean distributions. While noise is additive, fading is multiplicative, it scales the amplitude of the signal. But fading is not always bad, when multiple antennas are used, fading can be exploited to increase the capacity of the wireless channel.

Another important phenomenon is interference, which could be cochannel or adjacent channel. When the signal from a distant cell that uses the same frequency, interferes with the signal of interest, it is called cochannel interference. This can be avoided by careful frequency planning and/or by using directional antennas. More recently beamforming has been used to transmit to the mobile user directly without interfering with other unintended users in the vicinity. Power control can also be used to reduce cochannel interference. Finally, there is also adjacent channel interference which is caused by spillage of wireless signal in the neighboring channel or band. This

Y. Ahmed, *Recipes for Communication and Signal Processing*,
Signals and Communication Technology,
https://doi.org/10.1007/978-981-99-2917-7_1

can be avoided by using precisely designed filters and/or selecting those modulations that have lower sidelobes.

Diffraction is an effect that allows signals to bend around objects and reach places which are otherwise unreachable through line of sight propagation. One popular diffraction model is the knife edge diffraction model, which we discuss later in the chapter. Scattering is a phenomenon where a signal reflected from a rough surface disperses along the angle of reflection. The level of scattering is dependent on the size of the irregularities on the object relative to the wavelength of the signal. Lastly, there is distance-dependent path loss that tells us that in the Free Space—Line of Sight (LOS) environment power falls off as squared of the distance. But this can change in urban environments where power falls off as the fourth power of the distance or even higher. Some of the popular path loss models are Hata Model, Okumura Model, Stanford University Interim (SUI) Model, etc.

Fading is usually termed as a small-scale effect, whereas path loss is termed as a large-scale effect. The former results in signal fluctuations when a mobile user moves for a couple of meters, or even when he is stationary, but the environment is changing. The latter results in increase or decrease of received power when a user moves over a distance of 10 or 100 s of meters.

1.2 Noise Calibration in Simulation of Communication Systems

We will be using a wireless signal model in our simulations, in the next chapter, on Modulation and Coding. In this article we give a brief introduction to it. Let's assume the received signal is given as

$$r(t) = s(t) + n(t),$$

where $r(t)$ is the received signal and $s(t)$ is the transmitted signal and $n(t)$ is the Additive White Gaussian Noise (AWGN). We do not go into a lengthy discussion about fading model but give it here for completeness sake.

$$r(t) = h(t)s(t) + n(t)$$

Note that channel $h(t)$ is multiplicative, whereas noise $n(t)$ is additive. This is only true for very basic single tap channel models. Typically $h(t)$ has a Rayleigh distributed amplitude and uniformly distributed phase.

Signal to noise ratio (SNR) for the simulation of digital communication systems is given as

$$\rho = \frac{E_b}{N_o},$$

where E_b is the energy per bit and N_o is the noise power spectral density (PSD). We also know that for the case of Additive White Gaussian Noise the noise power is given as [1]

$$\sigma^2 = \frac{N_o}{2} f_s$$
$$\frac{2\sigma^2}{f_s} = N_o$$

Where σ is the standard deviation of noise and f_s is the sampling frequency. Substituting in the above equation we get

$$\rho = \frac{E_b}{N_o} = \frac{E_b f_s}{2\sigma^2}$$
$$\sigma^2 = \frac{E_b f_s}{2E_b/N_o}$$
$$\sigma = \sqrt{\frac{E_b f_s}{2E_b/N_o}}.$$

If the energy per bit and the sampling frequency is set to 1, the above equation reduces to: (Fig. 1.1)

$$\sigma = \sqrt{\frac{1}{2E_b/N_o}}.$$

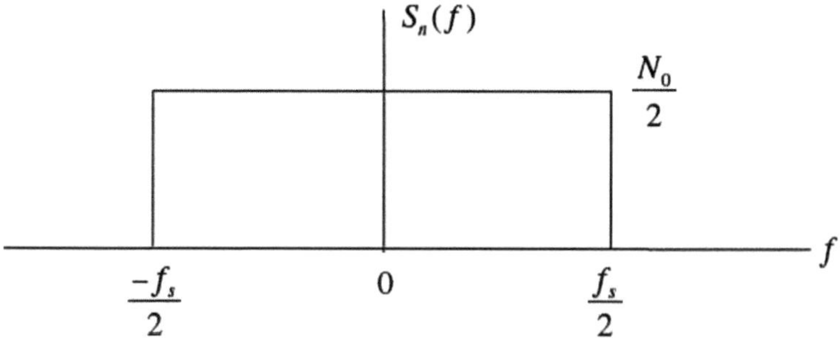

Fig. 1.1 Power spectral density (PSD) of noise

The simulation software can thus calculate the noise standard deviation (or variance) for each value of E_b/N_o in the simulation cycle. The following piece of MATLAB code generates AWGN with the required power and adds it to the transmitted signal.

```
s=sign(rand-0.5);          % Generate a symbol
sigma=1/sqrt(2*EbNo);      % Calculate noise standard deviation
n=sigma*randn;             % Generate AWGN with the required std dev
r=s+n;                     % Add noise to the signal
```

How can we assume that energy per bit and sampling frequency is equal to one and are we breaking some discrete time signal processing rule here? This is left as a point to ponder on.

1.3 Sum of Sinusoids Fading Simulator

There are various types of fading simulators that have been suggested in the literature, such as frequency domain fading simulators, i.e., simulators that define the Doppler components in the frequency domain and then perform an IDFT to get the time domain signal. These simulators include Smith's simulator, Young's simulator and our very own computationally efficient Rayleigh fading simulator [2]. Another technique that has been widely reported in the literature is Sum of Sinusoids Method. As the name suggests this method generates the Doppler components in the time domain and then sums them up to generate the time domain fading envelope. There are three parameters that define the properties of the generated signal.

(1) Number of sinusoids—Higher the number better the properties of the generated signal but greater the computational complexity
(2) Angle of arrival—This can be generated statistically or deterministically, spread from –pi to pi
(3) Phase of the arriving wave—This is uniformly distributed between –pi and pi

The MATLAB code below gives three similar sum of sinusoids techniques for generating a Rayleigh faded envelope [3].

```
%%%%%%%%%%%%%%%%%%%%%%%%%%%%%%%%%%%%%%%%%%%%%%%%%%%
%                  SUM OF SINUSOIDS FADING SIMULATORS
%
%                         fd is the doppler frequency
%                         fs is the sampling frequency
%                         ts is the sampling period
%                         N is the number of sinusoids
%
%                         Copyright 2020 RAYmaps
%%%%%%%%%%%%%%%%%%%%%%%%%%%%%%%%%%%%%%%%%%%%%%%%%%%
clear all
close all

fd=70;
fs=1000000;
ts=1/fs;
t=0:ts:1;
N=100;

%%%%%%%%%%%%%%%%%%%%%%%%%%%%%%%%%%%%%%%%%%%%%%%%%%%
% Method 1 - Clarke
%%%%%%%%%%%%%%%%%%%%%%%%%%%%%%%%%%%%%%%%%%%%%%%%%%%
x=zeros(1,length(t));
y=zeros(1,length(t));


for n=1:N;n
  alpha=(rand-0.5)*2*pi;
  phi=(rand-0.5)*2*pi;
  x=x+randn*cos(2*pi*fd*t*cos(alpha)+phi);
  y=y+randn*sin(2*pi*fd*t*cos(alpha)+phi);
end
z=(1/sqrt(N))*(x+1i*y);
r1=abs(z);
plot(t,10*log10(r1))
hold on

%%%%%%%%%%%%%%%%%%%%%%%%%%%%%%%%%%%%%%%%%%%%%%%%%%%
% Method 2 - Pop, Beaulieu
%%%%%%%%%%%%%%%%%%%%%%%%%%%%%%%%%%%%%%%%%%%%%%%%%%%
x=zeros(1,length(t));
y=zeros(1,length(t));
for n=1:N;n
  alpha=2*pi*n/N;
  phi=(rand-0.5)*2*pi;
  x=x+randn*cos(2*pi*fd*t*cos(alpha)+phi);
  y=y+randn*sin(2*pi*fd*t*cos(alpha)+phi);
end
z=(1/sqrt(N))*(x+1i*y);
r2=abs(z);
plot(t,10*log10(r2),'r')
hold on
```

```
%%%%%%%%%%%%%%%%%%%%%%%%%%%%%%%%%%%%%%%%%%%%%%%
% Method 3 - Chengshan Xiao
%%%%%%%%%%%%%%%%%%%%%%%%%%%%%%%%%%%%%%%%%%%%%%%
x=zeros(1,length(t));
y=zeros(1,length(t));
for n=1:N;n
  phi=(rand-0.5)*2*pi;
  theta=(rand-0.5)*2*pi;
  alpha=(2*pi*n+theta)/N;
  x=x+randn*cos(2*pi*fd*t*cos(alpha)+phi);
  y=y+randn*sin(2*pi*fd*t*cos(alpha)+phi);
end
z=(1/sqrt(N))*(x+1i*y);
r3=abs(z);
plot(t,10*log10(r3),'g')
hold off
xlabel('Time(sec)')
ylabel('Envelope(dB)')
%%%%%%%%%%%%%%%%%%%%%%%%%%%%%%%%%%%%%%%%%%%%%%%
```

All the three techniques given above are quite accurate in generating a Rayleigh faded envelope with the desired statistical properties. The accuracy of these techniques increases as the number of sinusoids goes to infinity (we have tested these techniques with up to 1000 sinusoids but realistically speaking even 100 sinusoids are enough). If we want to compare the three techniques in terms of the level crossing rate (LCR) and average fade duration (AFD) we can say that the first and third technique are a bit more accurate than the second technique. Therefore, we can conclude that a statistically distributed angle of arrival is a better choice than a deterministically distributed angle of arrival. Also, if we look at the autocorrelation of the in-phase and quadrature components, we see that for the first and third cases we get a zero-order Bessel function of the first kind, whereas for the second case we get a somewhat different sequence which approximates the Bessel function with increasing accuracy as the number of sinusoids is increased (Fig. 1.2).

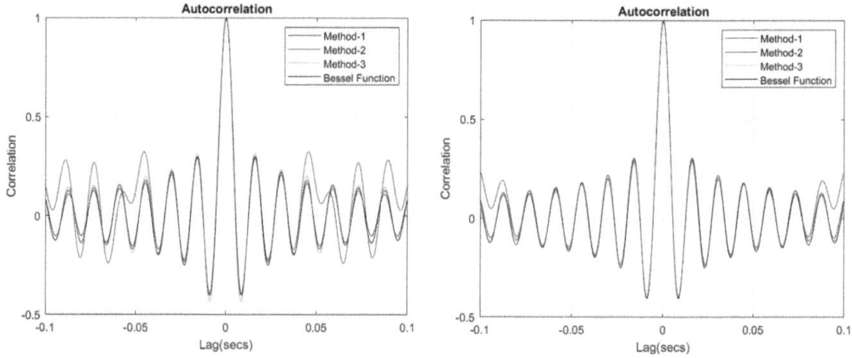

Fig. 1.2 Correlation of real and imaginary parts **a** 20 sinusoids **b** 40 sinusoids

Figure 1.2 shows the theoretical Bessel function versus the autocorrelation of the real/imaginary parts generated by the three methods. Figure on the left considers 20 sinusoids, whereas the figure on the right considers 40 sinusoids. As can be seen the accuracy of the autocorrelation sequence increases considerably by doubling the number of sinusoids. We can assume that for number of sinusoids exceeding 100, i.e., $N = 100$ in the above code the generated autocorrelation sequence would be quite accurate.

Note:

1. Significant averaging has been performed to generate the above curves, which results in an overlap with Bessel function (of first kind and order zero).
2. MATLAB function "xcorr" is used to generate the autocorrelation sequence and "besselj" is used to generate the Bessel function (J = besselj(0,2*pi*fd*tau)).

1.4 Correlated Rayleigh Fading Simulator

As discussed previously an LTE channel can be modeled as an FIR filter. The filter taps are described by the EPA, EVA and ETU channel models.

If $x(k)$ is the original signal, then the signal at the output of the FIR filter $y(k)$ is given as

$$y(k) = x(k)c(0) + x(k-1)c(1) + \cdots + x(k-L+2)c(L-2) + x(k-L+1)c(L-1)$$

Since the wireless channel is time varying the channel taps, $c(0), c(1) \ldots c(L-1)$ are also time varying with either Rayleigh or Rician distribution. It is quite easy to generate Rayleigh random variables with the desired power and distribution; however, when these Rayleigh random variables are required to have temporal correlation the process becomes a bit complicated. Temporal correlation of these variables depends upon the Doppler frequency which in turn depends upon the speed of the mobile device. The Doppler frequency is defined as: (Fig. 1.3).

$$f_d = v\ \cos(\theta)/\lambda$$

where

f_d is the Doppler Frequency in Hz

v is the receiver velocity in m/sec

λ is the wavelength in m

and θ is the angle between the direction of arrival of the signal and the direction of motion

A simple method for generating Rayleigh random variables with the desired temporal correlation was devised by Smith [4]. His method was based on Clark

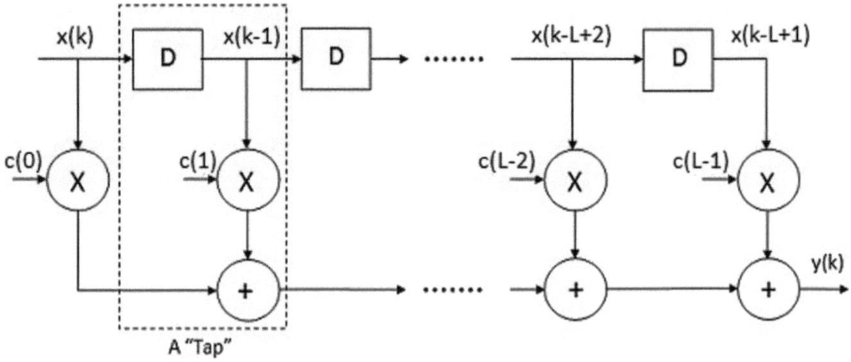

Fig. 1.3 Wireless propagation channel modeled as an FIR filter

and Gans fading model and has been widely used in simulation of wireless communication systems.

The method for generating the Rayleigh fading envelope with the desired temporal correlation is given below (modified from Theodore S. Rappaport Text).

1. Define N the number of Gaussian RVs to be generated, f_m the Doppler frequency in Hz, f_s the sampling frequency in Hz, d_f the frequency spacing which is calculated as $d_f = 2f_m/(N-1)$ Hz and M total number of samples in frequency domain which is calculated as $M = f_s/d_f$.
2. Generate two sequences of $N/2$ complex Gaussian random variables. These correspond to the frequency bins up to f_m. Take the complex conjugate of these sequences to generate the $N/2$ complex Gaussian random variables for the negative frequency bins up to $-f_m$.
3. Multiply the above complex Gaussian sequences g_1 and g_2 with square root of the Doppler Spectrum S generated from $-f_m$ to f_m. Calculate the spectrum at $-f_m$ and f_m by using linear extrapolation.
4. Extend the above-generated spectra from $-f_s/2$ to $f_s/2$ by stuffing zeros from $-f_s/2$ to $-f_m$ and f_m to $f_s/2$. Take the IFFT of the resulting spectra X and Y resulting in time domain signals x and y.
5. Add the absolute values of the resulting signals x and y in quadrature. Take the absolute value of this complex signal. This is the desired Rayleigh distributed envelope with the required temporal correlation.

The MATLAB code for generating Rayleigh random sequence with a Doppler frequency of f_m Hz is given below.

```
%%%%%%%%%%%%%%%%%%%%%%%%%%%%%%%%%%%%%%%%%%%%%%%%%%%%%%%%%%%%%%%%
%           RAYLEIGH FADING SIMULATOR BASED UPON SMITH'S METHOD
%                         N is the number of paths
%           M is the total number of points in the frequency domain
%                   fm is the doppler frequency in Hz
%                   fs is the sampling frequency in Hz
%               df is the step size in the frequency domain
%                       Copyright 2020 RAYmaps
%%%%%%%%%%%%%%%%%%%%%%%%%%%%%%%%%%%%%%%%%%%%%%%%%%%%%%%%%%%%%%%%
clear all;
close all;

N=1000;
fm=300;
df=(2*fm)/(N-1);
fs=7.68e6;
M=round(fs/df);
T=1/df;
Ts=1/fs;

% Generate two sequences of N complex Gaussian random variables
g=randn(1,N/2)+j*randn(1,N/2);
gc=conj(g);
g1=[fliplr(gc), g];

g=randn(1,N/2)+j*randn(1,N/2);
gc=conj(g);
g2=[fliplr(gc), g];

% Generate Doppler Spectrum S
f=-fm:df:fm;
S=1.5./(pi*fm*sqrt(1-(f/fm).^2));
S(1)=2*S(2)-S(3);
S(end)=2*S(end-1)-S(end-2);

% Multiply the sequences with the Doppler Spectrum S, take IFFT
X=g1.*sqrt(S);
X=[zeros(1,round((M-N)/2)), X, zeros(1,round((M-N)/2))];
x=abs(ifft(X,M));

Y=g2.*sqrt(S);
Y=[zeros(1,round((M-N)/2)), Y, zeros(1,round((M-N)/2))];
y=abs(ifft(Y,M));

% Find the resulting Rayleigh faded envelope
z=x+j*y;
r=abs(z);
r=r/sqrt(mean(r.^2));
t=0:Ts:T-Ts;
plot(t,10*log10(r),'r')
xlabel('Time(sec)')
ylabel('Envelope (dB)')
axis([0 0.01 -10 5])
%%%%%%%%%%%%%%%%%%%%%%%%%%%%%%%%%%%%%%%%%%%%%%%%%%%%%%%%%%%%%%%%%%%%%%
```

The above code generates a Rayleigh random sequence with samples spaced at 0.1302 usec. This corresponds to a sampling frequency of 7.68 MHz which is the standard sampling frequency for a bandwidth of 5 MHz. Similarly, the sampling rate for 10 MHz and 20 MHz is 15.36 MHz and 30.72 MHz, respectively. The Doppler frequency can also be changed according to the scenario. LTE standard defines 3 channel models EPA, EVA and ETU with Doppler frequencies of 5 Hz, 70 Hz and 300 Hz, respectively. These are also known as Low Doppler, Medium Doppler and High Doppler, respectively.

The above code generated Rayleigh sequences of varying lengths for the three cases. But in all the cases it is in excess of 10 ms and can be used as the fading sequence for an LTE frame. Just to recall an LTE frame is of 10 ms duration with 20 time slots of 0.5 ms each. If each slot contains 7 OFDM symbols, the total length of a fading sequence is 140 symbols (Fig. 1.4).

It is seen that the fluctuation in the channel increases with Doppler frequency. The channel is almost static for a Doppler frequency of 5 Hz and varies quite rapidly for a Doppler frequency of 300 Hz. It is also shown above that the envelope of z is Rayleigh distributed and phase of z is uniformly distributed. However, the range of phase is from 0 to pi/2. This needs to be further investigated. The level crossing rate and average fade duration can also be measured (Fig. 1.5).

This is the process for generating one Rayleigh distributed channel tap. This step would have to be repeated for the number of taps in the channel model which could be 7 or 9 for the LTE channel models. A Ricean distributed channel tap can be generated in a similar fashion. MIMO channel taps can also be generated using

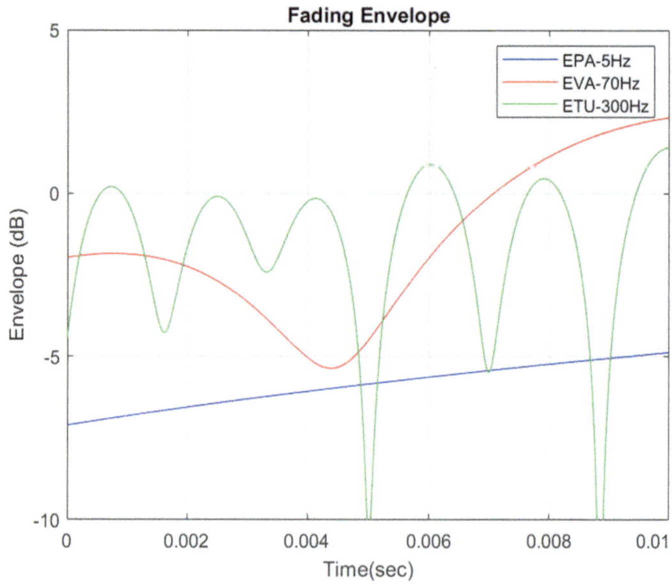

Fig. 1.4 Temporal variations of the three wireless channel (EPA, EVA, ETU)

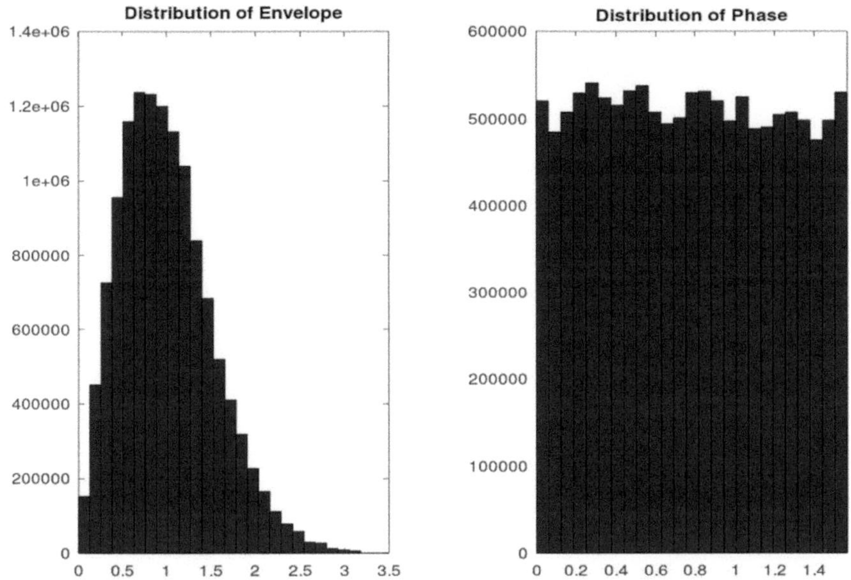

Fig. 1.5 Rayleigh fading envelope and phase distribution

the above-described method; however, we would need to understand the concept of antenna correlation before we do that.

Level Crossing Rate and Average Fade Duration

Level crossing rate (LCR) is defined as number of times per second the signal envelope crosses a given threshold. This could be either in the positive direction or negative direction. Average fade duration (AFD) is the average duration that the signal envelope remains below a given threshold once it crosses that threshold. Simply it is the average duration of a fading event. The LCR and AFD are interconnected and the product of these two quantities is a constant. The program given below calculates the LCR and AFD of the above-generated envelope $r(t)$.

The program calculates both the simulated and theoretical values of LCR and AFD, e.g., for a threshold level of 0.3 (-5.22 dB) the LCR and AFD values calculated for $f_m = 70$Hz and $N = 32$ are:

LCR simulation = 45.16
LCR theoretical = 43.41
AFD simulation = 0.0015 s
AFD theoretical = 0.0016 s

It can be seen that the theoretical and simulation results match quite well. This gives us confidence that is generated envelope has the desired statistical characteristics.

```
%%%%%%%%%%%%%%%%%%%%%%%%%%%%%%%%%%%%%%%%%%%%%%%%%%%%%
%                    PROGRAM TO CALCULATE THE LCR and AFD
%                  Rth is the threshold to calculate the LCR and AFD
%                         Rrms is the RMS level of the signal r
%                  rho is the ratio of defined threshold and RMS level
%                               Copyright 2020 RAYmaps
%%%%%%%%%%%%%%%%%%%%%%%%%%%%%%%%%%%%%%%%%%%%%%%%%%%%%
Rth=0.30;
Rrms=sqrt(mean(r.^2));
rho=Rth/Rrms;

count1=0;
count2=0;
for n=1:length(r)-1
  if r(n) < Rth && r(n+1) > Rth
    count1=count1+1;
  end
  if r(n) < Rth
    count2=count2+1;
  end
end
LCR=count1/(T)
AFD=((count2*Ts)/T)/LCR

LCR_num=sqrt(2*pi)*fm*rho*exp(-(rho^2))
AFD_num=(exp(rho^2)-1)/(rho*fm*sqrt(2*pi))
%%%%%%%%%%%%%%%%%%%%%%%%%%%%%%%%%%%%%%%%%%%%%%%%%%%%%
```

Note:

1. According to Wireless Communications Principles and Practice by Ted Rappaport "Perform an IFFT on the resulting frequency domain signals from the in-phase and quadrature arms to get two N-length time series, and add the squares of each signal point in time to create an N-point time series. Note that each quadrature arm should be a real signal after IFFT." Now this point about the signal being real after IFFT is not always satisfied by the above program. The condition can be satisfied by playing around with the value of N a bit.
2. Also, we take the absolute value of both the time series after IFFT operation to make sure that we get a real-valued sequence. However, taking the absolute value of both the in-phase and quadrature terms make z fall in the first quadrant and the phase of z to vary from 0 to pi/2. A better approach might be to use the "real" function instead of "abs" function so that the phase can vary from 0 to 2 pi.
3. A computationally efficient method of generating Rayleigh fading sequence is given here.

1.5 Knife Edge Diffraction Model

Diffraction is a phenomenon where electromagnetic waves (such as light waves) bend around corners to reach places which are otherwise not reachable, i.e., not in the line of sight. In technical jargon such regions are also called shadowed regions

Fig. 1.6 Rays being regenerated after coming in interaction with an object

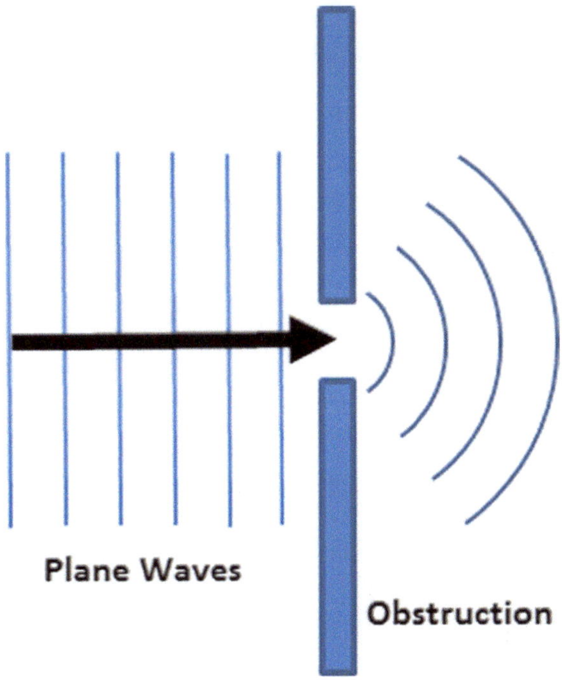

(the term again drawn from the physics of light). This phenomenon can be explained by Huygen's principle which states that "as a plane wave propagates in a particular direction each new point along the wavefront is a source of secondary waves." This can be understood by looking at the above figure. However, one peculiarity of this principle is that it is unable to explain why the new point source transmits only in the forward direction (Fig. 1.6).

The electromagnetic field in the shadowed region can be calculated by combining vectorially the contributions of all of these secondary sources, which is not an easy task. Secondly, the geometry is usually much more complicated than shown in the figure. For example, consider a telecom tower transmitting electromagnetic waves from a rooftop and a pedestrian using a mobile phone at street level. The EM waves usually reach the receiver at street level after more than one diffraction (not to mention multiple reflections). However, an approximation that works well in most cases is called knife edge diffraction, which assumes a single sharp edge (an edge with a thickness much smaller than the wavelength) separates the transmitter and receiver.

The path loss due to diffraction in the knife edge model is controlled by the Fresnel Diffraction Parameter which measures how deep the receiver is within the shadowed region. A negative value for the parameter shows that the obstruction is below the line of sight and if the value is below -1 there is hardly any loss. A value of 0 (zero) means that the transmitter, receiver and tip of the obstruction are all in line and the Electric Field Strength is reduced by half or the power is reduced to one-fourth of the value without the obstruction, i.e., a loss of 6 dB. As the value of the Fresnel

Diffraction Parameter increases on the positive side the path loss rapidly increases reaching a value of 27 dB for a parameter value of 5. Sometimes the exact calculation is not needed and only an approximate calculation, as proposed by Lee in 1985, is sufficient. Fresnel Diffraction Parameter v is defined as:

$$v = h\sqrt{\frac{2(d_1 + d_2)}{\lambda d_1 d_2}}$$

d_1 is the distance between the transmitter and the obstruction along the line of sight
d_2 is the distance between the receiver and the obstruction along the line of sight
h is the height of the obstruction above the line of sight
and λ is the wavelength

The electrical length of the path difference between a diffracted ray and a LOS ray is equal to $\varphi = (\pi/2)v^2$ and the normalized electric field produced at the receiver, relative to the LOS path is $e^{-j\varphi}$. Performing a summation of all the exponentials above the obstruction (from v to positive infinity) gives us the Fresnel Integral, $F(v)$ (Figs. 1.7 and 1.8).

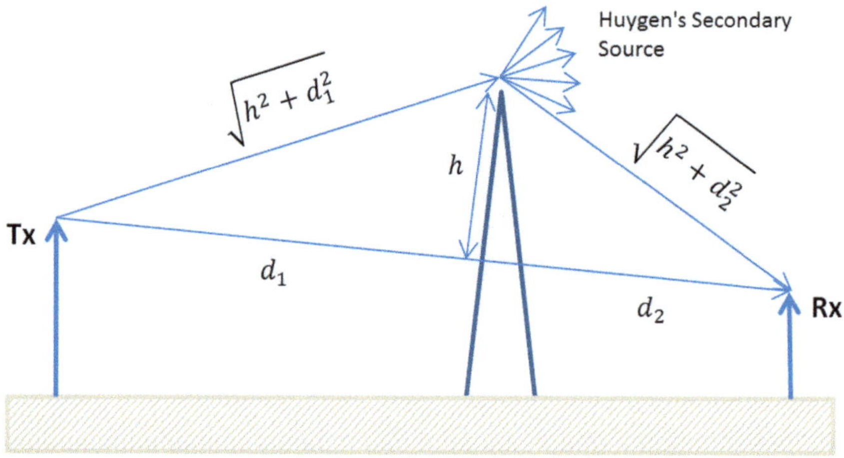

Fig. 1.7 Knife edge diffraction model using Huygen's principle

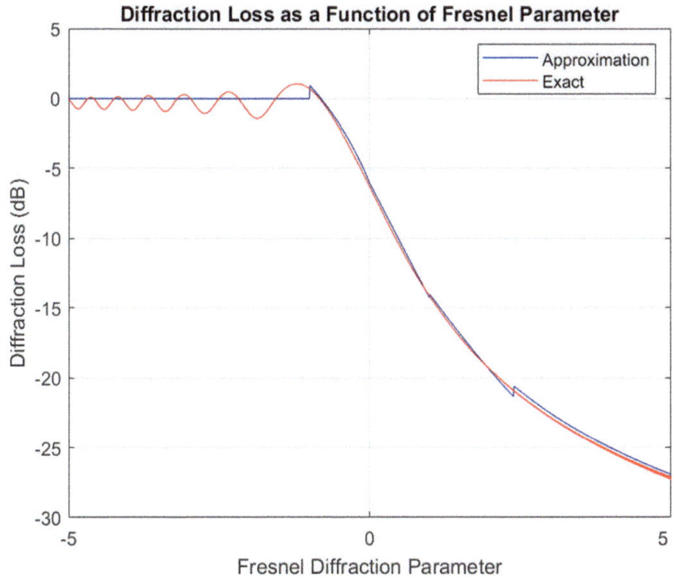

Fig. 1.8 Diffraction loss as a function of Fresnel diffraction parameter

The MATLAB codes used to generate the above plots are given below (approximate method followed by the exact method). Feel free to use them in your simulations and if you have a question drop us a comment.

```
%%%%%%%%%%%%%%%%%%%%%%%%%%%%%%%%%%%%%%%%%%%%%%%%%%%%%%%%%%%%%
%                  Calculation of the path loss based on the value of
%                   Fresnel Diffraction Parameter as proposed by Lee
%                 Lee W C Y Mobile Communications Engineering 1985
%
%                         v is the Fresnel Diffraction Parameter
%                              G is the diffraction loss in dB
%                                Copyright 2020 RAYmaps
%%%%%%%%%%%%%%%%%%%%%%%%%%%%%%%%%%%%%%%%%%%%%%%%%%%%%%%%%%%%%
clear all
close all

v=-5:0.01:5;

for n=1:length(v)

  if v(n) <= -1
    G(n)=0;
  elseif v(n) <= 0
    G(n)=20*log10(0.5-0.62*v(n));
  elseif v(n) <= 1
    G(n)=20*log10(0.5*exp(-0.95*v(n)));
  elseif v(n) <= 2.4
    G(n)=20*log10(0.4-sqrt(0.1184-(0.38-0.1*v(n))^2));
  else
    G(n)=20*log10(0.225/v(n));
  end

end
plot(v, G, 'b')
xlabel('Fresnel Diffraction Parameter')
ylabel('Diffraction Loss (dB)')
```

```
%%%%%%%%%%%%%%%%%%%%%%%%%%%%%%%%%%%%%%%%%%%%%%%%
%              Exact calculation of the path loss (dB)
%              based on Fresnel Diffraction Parameter (v)
%              T S Rappaport Wireless Communications P&P
%              v is the Fresnel Diffraction Parameter
%              G is the diffraction loss in dB
%              Copyright 2020 RAYmaps
%%%%%%%%%%%%%%%%%%%%%%%%%%%%%%%%%%%%%%%%%%%%%%%%
clear all
close all
v=-5:0.01:5;

for n=1:length(v)
  v_vector=v(n):0.01:v(n)+100;
  F(n)=((1+1i)/2)*sum(exp((-1i*pi*(v_vector).^2)/2));
end

F=abs(F)/(abs(F(1)));
plot(v, 20*log10(F),'r')
xlabel('Fresnel Diffraction Parameter')
ylabel('Diffraction Loss (dB)')
%%%%%%%%%%%%%%%%%%%%%%%%%%%%%%%%%%%%%%%%%%%%%%%%
```

We have used the following equations in the exact calculation of the diffraction loss [5] above. We did not want to scare you with the math so have saved it for the end.

$$\frac{E_d}{E_{LOS}} = F(v) = \frac{1+j}{2}\int_v^{\infty} e^{-j\frac{\pi}{2}t^2}dt$$

$$F(v) = \frac{1+j}{2}\left[\int_v^{\infty}\cos\left(\frac{\pi}{2}t^2\right)dt - j\int_v^{\infty}\sin\left(\frac{\pi}{2}t^2\right)dt\right]$$

$$v = h\sqrt{\frac{2(d_1+d_2)}{\lambda d_1 d_2}}$$

Also please checkout this interesting video explaining the phenomenon of diffraction.

1.6 Simulating a SISO Ring Model

A Ring Model is a well-known spatial channel model also known as a geometrical model. It models the propagation channel as an unobstructed transmitter and a receiver surrounded by a ring of reflectors. The distance between the transmitter and receiver is usually much larger than the radius of the ring. The reflectors are distributed uniformly around the ring. This model is useful for modeling a scenario

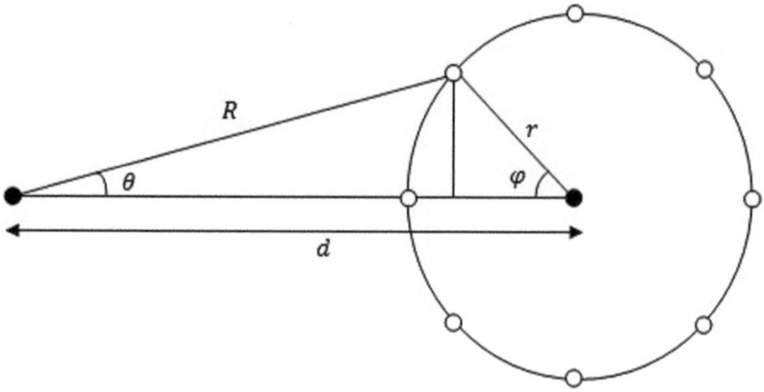

$$R = \sqrt{(d - r\cos\varphi)^2 + (r\sin\varphi)^2}$$

Fig. 1.9 Geometry of the SISO ring model

where a base station is located at sufficient altitude and is unobstructed, whereas the mobile station is at ground level and is surrounded by a bunch of reflectors (Fig. 1.9).

We simulate the SISO Ring Model described previously by varying the transmit receive separation from 0 to 500 m. Keeping the ring radius fixed at 20 m the angular spread of the channel decreases as the receiver moves away from the transmitter. Frequency is assumed to be 1 GHz, and phase constant is calculated as 20.94 radians/m. It is obvious that when the distance is small, particularly less than the radius of the ring, there is a lot of constructive and destructive interference. But this is only shown here for purpose of completeness, in reality distance is always much greater than radius of the ring. It would be advisable to look at the variations of power for a distance in excess of 100 m.

Given below is the MATLAB code and the plot for a SISO Ring Model consisting of eight reflectors distributed uniformly around the ring.

It is observed that the power level of the received signal fluctuates as the distance d is varied. However, after a certain critical distance (around 250 m) the power versus distance curve approaches a straight line (almost) (Fig. 1.10).

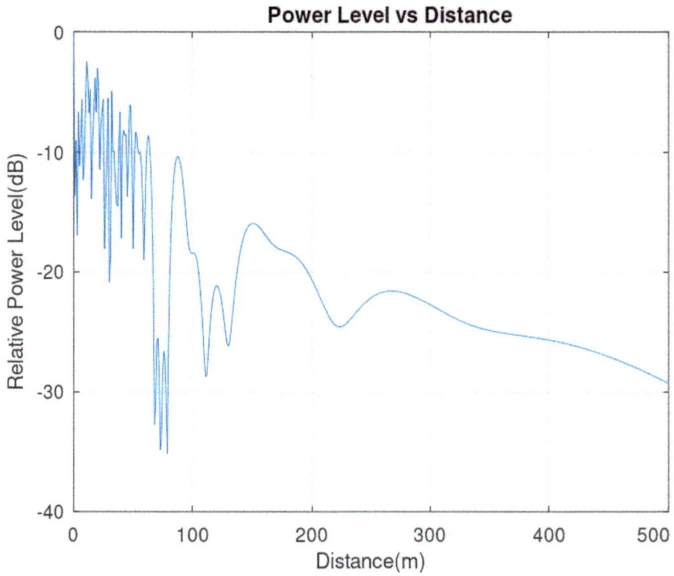

Fig. 1.10 Simulating a SISO ring model

```
%%%%%%%%%%%%%%%%%%%%%%%%%%%%%%%%%%%%%%%%%%%%%%%%%%
%                         SISO RING MODEL
%          d is the separation between transmitter and receiver
%                      r is the radius of the ring
%                      B is the phase constant
%                      Et is the resultant E-field
%                     Copyright 2020 RAYmaps
%%%%%%%%%%%%%%%%%%%%%%%%%%%%%%%%%%%%%%%%%%%%%%%%%%
function [Et]=ring_mod0(d,r,B)
Et=0;
for phi=0:pi/4:(2*pi)-(pi/4);
  R=sqrt((d-r*cos(phi))^2+(r*sin(phi))^2);
  E=exp(-i*B*(r+R))/(r+R);
  Et=Et+E;
end
return
%%%%%%%%%%%%%%%%%%%%%%%%%%%%%%%%%%%%%%%%%%%%%%%%%%
```

Note:

1. Extending the SISO Ring Model to MIMO Ring Model is not that difficult, and it is left to the reader to experiment with.
2. It is very important to select the right phase constant for the scenario under study as constructive and destructive interference depends upon the value of the phase constant (which depends on the wavelength).

3. It is customary to plot power instead of E-field with respect to distance in such scenarios.

1.7 Near Field and Far Field of an Antenna

The electromagnetic radiation from an antenna, particularly dipole antenna, has been studied in great detail. The mathematical framework proposed by Maxwell has stood the test of time and theoretical concepts have been verified through physical measurements. But the behavior of electromagnetic (EM) waves close to the radiating antenna is not that well understood. This region that extends to about a wavelength from the antenna is called near field, as opposed to far field, which extends further out. The near field is further divided into reactive near field and radiative near field [6].

We know that in the far field, the E-field, H-field and directional of propagation are all perpendicular to each other and E-field and H-field are in phase. The impedance of free space (given as ratio of E-field and H-field) in this region is equal to 377 ohms (120π ohms). But this is not the case in near field where E-field is much greater than H-field, and these fields are not in phase. Furthermore the rate of decay of E-field and H-field is much higher in the near field (Fig. 1.11).

Given below is the code that plots the E-field and H-field (normalized, so that rate of decay can be compared) in the three regions. It can be seen that there is no clear

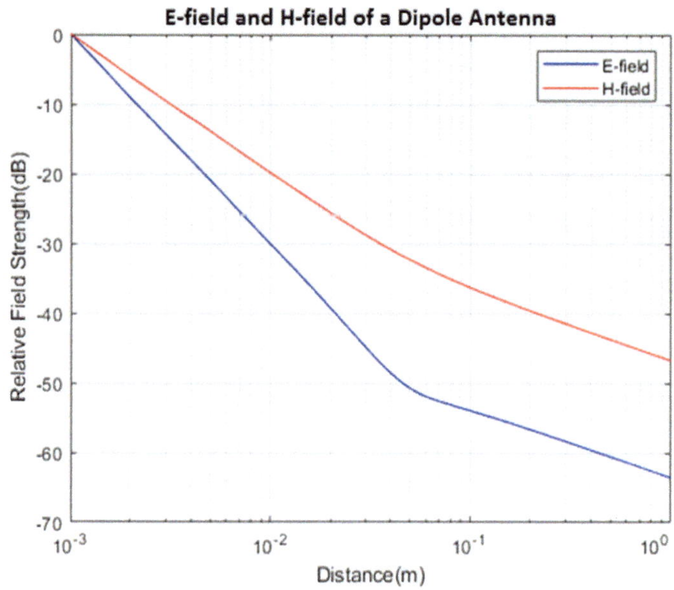

Fig. 1.11 E-field and H-field of a small dipole antenna

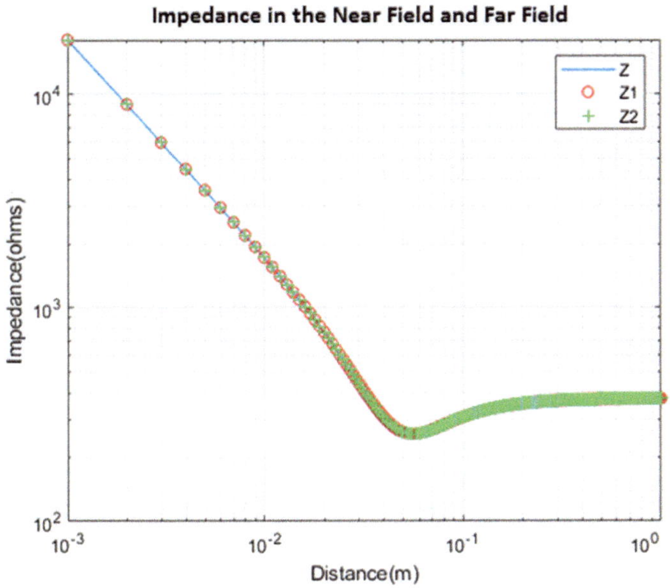

Fig. 1.12 Impedance in the near field and far field

boundary defining the three regions; the fields and resulting impedance gradually change as the distance from the transmit antenna increases. In the far field both the E-field and H-field decay at 10 dB/decade. In the region closest to the antenna E-field decays at 30 dB/decade, whereas H-field decays as 20 dB decade (Fig. 1.11).

Note:

1. Please note that we have used the E-field and H-field equations derived for a small dipole and kept the length of the dipole to be one-tenth of the wavelength.
2. We have used three methods to plot the impedance; from the definition of impedance ($Z = E/H$), from the formula for impedance (Z_1) and simplified formula for impedance (Z_2).
3. The exact formula for calculating the length of near field is given as $d_{NF} = 2D^2/\lambda$, where λ is the wavelength and D is the length of the antenna. But this formula is only valid for large antennas.

```
%%%%%%%%%%%%%%%%%%%%%%%%%%%%%%%%%%%%%%%%%%%%%%%%%%%%%%%
%                    E-field and H-field of a Small Dipole Antenna
%                              f is the frequency
%                              c is the speed of light
%                              lambda is the wavelength
%                              k is the wave number
%                              L is the length of the antenna
%                    nue is the characteristic impedance of free space
%
%                              Copyright 2020 RAYmaps
%%%%%%%%%%%%%%%%%%%%%%%%%%%%%%%%%%%%%%%%%%%%%%%%%%%%%%%
clear all
close all

f=1e9;
c=3e8;
lambda=c/f;
k=2*pi/lambda;
L=lambda/10;
nue=377;
Io=1;
theta=pi/2;

r=0.001:0.001:1;

E_theta=i*nue*k*Io*L*sin(theta)./(4*pi*r).*(1+1./(i*k*r)+1./(i*k*r).^2).*exp(-i*k*r);
semilogx(r,10*log10(abs(E_theta)/max(abs(E_theta))),'b');hold on
H_phi=i*k*Io*L./(4*pi*r).*(1+1./(i*k*r))*sin(theta).*exp(-i*k*r);
semilogx(r,10*log10(abs(H_phi)/max(abs(H_phi))),'r');hold off
xlabel('Distance(m)')

ylabel('Relative Field Strength(dB)')
legend('E-field','H-field')
grid on

Z=E_theta./H_phi;
figure;loglog(r,abs(Z));hold on

Z1=nue*(1+1./(i*k*r)+1./(i*k*r).^2)./(1+1./(i*k*r));
loglog(r,abs(Z1),'ro');hold on

Z2=nue*(1-1./(k^2*r.^2-i*k*r));
loglog(r,abs(Z2),'g+');hold off
xlabel('Distance(m)')
ylabel('Impedance(ohms)')
legend('Z','Z1','Z2')
grid on
%%%%%%%%%%%%%%%%%%%%%%%%%%%%%%%%%%%%%%%%%%%%%%%%%%%%%%%
```

Questions and Numerical Problems

1. How long would it take for an electromagnetic wave to travel around the earth? You can assume that the radius of the earth is 6400 km.
2. What is the delay experienced by a wireless system that transmits a signal from a ground station to a GEO satellite (at 36,000 km) which then retransmits it to a dish antenna on the ground? Assume that there is no processing delay.
3. If the frequency of a wireless system is doubled what happens to the (a) Wavelength (b) Received power in free space line of sight, keeping all the other variables constant?
4. In free space line of sight conditions, if the distance increases by a factor of ten then by what factor does received power change?
5. What is the typical path loss exponent in (a) Free space line of sight environment (b) Typical urban environment?
6. If gain of the transmit antenna is 20 dB and gain of the receive antenna is also 20 dB, what is the total gain of the system in dB?
7. If the gain of the transmit antenna is 100 and gain of the receive antenna is also 100, what is the total gain on (a) Linear scale (b) Logarithmic scale?
8. Define level crossing rate and average fade duration. What is their relationship?
9. If the Doppler frequency increases what happens to the temporal correlation of the fading envelope?
10. What is the relationship between energy per bit to noise power spectral density ratio and signal to noise ratio?
11. How far does the reactive near field extend from the antenna. How far does the radiative near field extend from the antenna?
12. What is the rate of decay of E-field and H-field in the (a) Near field (b) Far field?
13. What is the relationship between permittivity, permeability and impedance of free space?
14. Can diffraction be a useful phenomenon? Please explain how? Give examples.
15. Based on the theory of SISO Ring Model given earlier, derive the mathematical relationships for MIMO Ring Model. Simulate the received power as a function of distance.

Useful Links

1. Noise Calibration in Simulation of Communication Systems
https://www.raymaps.com/index.php/noise-calibration-in-simulation-of-communication-systems/
2. Sum of Sinusoids Fading Simulator
https://www.raymaps.com/index.php/sum-of-sinusoids-fading-simulator/
3. Correlated Rayleigh Fading Simulator
https://www.raymaps.com/index.php/lte-fading-simulator/

4. Knife Edge Diffraction Model
   https://www.raymaps.com/index.php/knife-edge-diffraction-model/
5. Simulating a SISO Ring Model
   https://www.raymaps.com/index.php/simulating-a-siso-ring-model/
6. Near Field and Far Field of an Antenna
   https://www.raymaps.com/index.php/near-field-of-an-antenna/

References

1. Tranter, W.H., Shanmugan, K., Rappaport, T.S., Kosbar, K.: In: Principles of Communication Systems Simulation With Wireless Applications, Prentice Hall Press (2003)
2. Ahmed, Y., Reed, J.H.: An efficient implementation of young's fading simulator. In: 12th International Conference on Frontiers of Information Technology, December 2014, pp. 144–148 (2014)
3. Xiao, C: Novel sum-of-sinusoids simulation models for rayleigh and Rician fading channels. IEEE Trans. Wireless Commun. **5**(12) (2006)
4. Smith, J.I.: A computer generated multipath fading simulation for mobile radio. IEEE Trans. Vehicul. Technol. **VT-24**(3) (1975)
5. http://www.waves.utoronto.ca
6. https://nptel.ac.in/content/storage2/courses/108101092/Week-2-Dipole%20Antenna.pdf

Chapter 2
Modulation and Coding

2.1 Introduction

Somebody might ask why modulation and channel coding are in the same chapter. My answer is that these are two essential components of all modern digital communication systems. These two together allow for high-speed, error-free transmission over a wireless channel which is prone to noise, fading and interference. A strong code allows us to work at a low signal to noise ratio (SNR), and a higher-order modulation ensures that more bits are packed in a limited time frame. Without a strong code, higher-order modulations such as 256-QAM or 1024-QAM would not be viable.

The codes that we discuss in this chapter are called forward error correction (FEC) codes that do not require a feedback channel or retransmission. FEC codes can both detect and correct errors provided that there is not a complete erasure of a block or frame. We discuss three types of codes, namely block codes, convolutional codes and low-density parity check (LDPC) codes. The first two are quite popular codes but the third one got lost in the literature until it resurfaced in 2000s. LDPC codes are somewhat similar to turbo codes as both can be iteratively decoded and approach the Shannon Capacity.

Now back to why we need modulation. Following are some of the reasons (this list is not exhaustive).

i. Upconversion from baseband to passband (you need to transmit at the allocated frequency such as 88–108 MHz for FM radio)
ii. Allowing practical antennas to be designed; antenna size is proportional to the wavelength e.g. a half wavelength dipole would be 15 cm long at 1 GHz and 150 m at 1 MHz
iii. Achieve higher spectral efficiency by using higher-order modulations

Y. Ahmed, *Recipes for Communication and Signal Processing*,
Signals and Communication Technology,
https://doi.org/10.1007/978-981-99-2917-7_2

iv. Control spillage into adjacent frequency bands such as by using MSK modulation and Gaussian filter in GMSK.

I would also like to point out that real signals are used in the real world but complex baseband equivalents are used in mathematical proofs and simulation as it makes the analysis much simpler. Imagine that BPSK modulation is used to achieve a data rate of 20 Mbps over a channel at 1 GHz. The carrier signal at 1 GHz has a time period of 1nsec, whereas the symbol or bit period here is 50 nsec. So we would need 50 cycles of the carrier or at least 100 samples per symbol in our simulation. In a complex baseband simulation we can use 1 sample per symbol and get completely valid results.

The complex baseband representation we typically use is $a + jb$. If $a = \pm 1$ and $b = 0$, we get BPSK modulation. If $a = \pm 1, \pm 3$ and $b = 0$, we get 4-PAM. If $a = \pm 1$ and $b = \pm 1$, we get QPSK or 4-QAM modulation. However, the case for MSK modulation is not that simple as we typically need multiple samples per symbol due to continuous change of phase (MSK is also called continuous phase modulation). A somewhat crude representation of MSK signal is $a = \pm 1, b = 0$ and $a = 0, b = \pm 1$ during alternative symbol periods. Unlike other modulations MSK modulation has memory and phase can only change by 90° from one symbol to the next. This means that MSK can be demodulated using the Viterbi algorithm, resulting in a very low bit error rate.

2.2 Binary Phase Shift Keying Bit Error Rate in AWGN

Modulation is the process by which a binary stream (zeros and ones) is converted to a format that is suitable for transmission over a wired or wireless channel that is prone to noise and interference as well as distortion. The most basic modulation scheme is BPSK or Binary Phase Shift Keying. It transmits the information in the phase of the signal which could be one of two values (0° or 180°).

BPSK signal can be represented as (called the passband representation)

$$s(t) = a(t) \cos(2\pi f t),$$

where $a(t)$ is a time-varying parameter which can have one of two values (+1 or -1). This is equivalent to having the phase of the carrier rotated by 0° or 180°. In the simulation of digital communications systems, we usually take out the carrier and perform the simulation at baseband. The passband and baseband simulations are equivalent because the carrier signal introduced at the transmitter can be easily removed at the receiver by a process called correlation (or simply put multiplication by the carrier followed by low-pass filtering) and what we are left with is the parameter $a(t)$. If the transmitted signal is given as

$$s(t) = a(t)\cos(2\pi f t),$$

then by multiplication with the carrier at the receiver we get

$$\begin{aligned} s(t)\cos(2\pi f t) &= a(t)\cos(2\pi f t)\cos(2\pi f t) \\ &= \frac{a(t)}{2}(1 + \cos(4\pi f t)) \end{aligned}$$

and after low-pass filtering the cosine term at twice the carrier frequency is removed and we get the parameter $a(t)$ scaled by the factor 1/2. Since the information is contained in the sign of the parameter $a(t)$, we can recover our transmitted symbols.

So, in simulation, instead of multiplying the parameter $a(t)$ by the carrier at transmitter and then again at the receiver we simply transmit $a(t)$. This is equivalent to simulation of a Pulse Amplitude Modulation (PAM) system with two levels. Following are the steps involved in the simulation of BPSK system.

Steps:

1. Generate a random sequence of symbols (+1, −1).
2. Generate samples of Additive White Gaussian Noise (AWGN) with the required variance (noise power = noise variance OR noise power = square of noise standard deviation OR noise power = noise power spectral density * signal bandwidth).
3. Add AWGN samples to the BPSK signal.
4. Detection is performed at the receiver by determining the sign of the parameter $a(t)$.
5. And finally the bit error rate (BER) is calculated. Which is the same as symbol error rate (SER) in this case.

Given below is the MATLAB code that performs these functions. Also shown below are the signals generated at the first four steps and the bit error rate calculated in the fifth and last step. The length of the symbol sequence and EbNo (a bit different than SNR) are the inputs to the function, and the bit error rate (BER) is the output. The length of the sequence must be such that you can count about 25 symbol errors at each value of EbNo. This means that at an EbNo of 10 dB you would need to pass a few million symbols through the channel. Try it out!!! (Fig. 2.1 and 2.2).

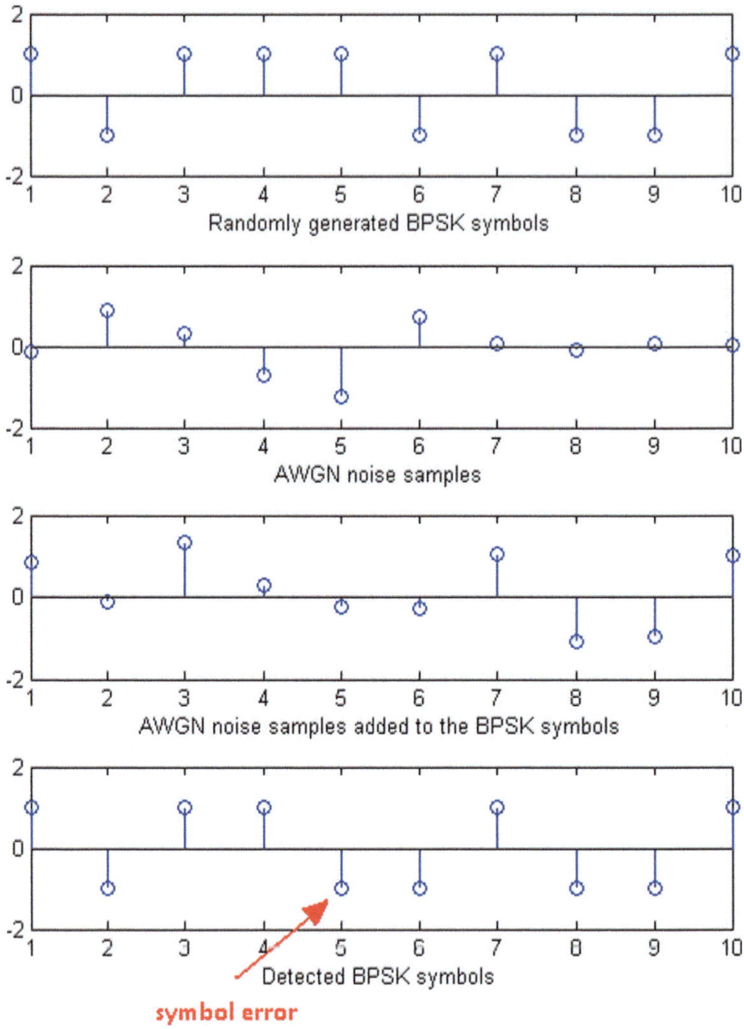

Fig. 2.1 BPSK symbols as they move the transmitter to the receiver

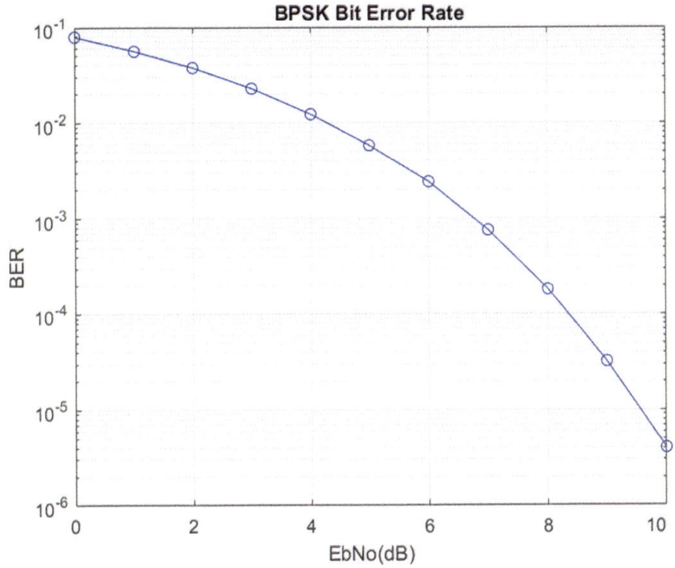

Fig. 2.2 Bit error rate of BPSK as a function of EbNo

```
%%%%%%%%%%%%%%%%%%%%%%%%%%%%%%%%%%%%%%%%%%%%%%%%%%%%%%%%%%%%%%%
%                 FUNCTION TO CALCULATE BER OF BPSK IN AWGN
%                      l is length of the symbol sequence
%             EbNo is energy per bit to noise power spectral density in dB
%                          ber is output bit error rate
%                          Copyright 2020 RAYmaps
%%%%%%%%%%%%%%%%%%%%%%%%%%%%%%%%%%%%%%%%%%%%%%%%%%%%%%%%%%%%%%%
function[ber]=err_rate(l,EbNo)
s=2*(round(rand(1,l))-0.5);
n=(1/sqrt(2*10^(EbNo/10)))*randn(1,l);
r=s+n;
s_=sign(r);
ber=(l-sum(s==s_))/l;
return
%%%%%%%%%%%%%%%%%%%%%%%%%%%%%%%%%%%%%%%%%%%%%%%%%%%%%%%%%%%%%%%
```

Note:

1. To generate the above-given bit error rate plot you would have to create a piece of code which calls the above function for each value of EbNo and stores the output BER value in an array and then plot the BER versus EbNo at the end of simulation. We leave this to you as an exercise.
2. We have generated BPSK symbols directly instead of first generating a binary sequence. This does not matter much in this simple example but for more advanced modulation schemes we would have to first generate a binary stream and then from that the symbols.

3. We have used one sample per symbol of BPSK modulation, as shown in the figure above. But sometimes we have to select higher number of samples per symbol (usually 4 to 10) to implement some other signal processing functions.
4. Most of the concepts discussed above can be extended to other digital modulation schemes. The concepts for analog modulation schemes (AM, FM, PM) are somewhat different and we do not use error rates to evaluate the performance of these schemes.

2.3 Pulse Amplitude Modulation Symbol Error Rate in AWGN

Pulse Amplitude Modulation (PAM) is a one dimensional or in other words real modulation. Simply put it is an extension of BPSK with M amplitude levels instead of two. This can be a bit confusing because BPSK can be looked at as a phase modulation, and its natural extension must be QPSK or 8-PSK modulations. To remove this ambiguity let's call M-PAM an extension of simple Amplitude Modulation but with M levels. In the discussion below we consider $M = 4$ but then extend it to the general case of $M = 2^k$ $(k = 1, 2, 3 \ldots)$.

The symbol generation and detection of 4-PAM is slightly more complicated than BPSK/BASK/OOK as four amplitude levels are involved. A "for loop" with four conditions are used for both symbol generation and detection. The four symbols are $\pm$ 1 and $\pm$ 3, and the average energy per symbol is calculated as $(1 + 1 + 9 + 9)/4 = 5$. Remember that the energy per symbol is simply the average of squared Euclidean distances of the symbols from the origin.

As is customary Additive White Gaussian Noise (AWGN) with standard deviation of sigma is added to the signal. This can be used to vary the energy per symbol to noise power spectral density ratio (EsNo). Theoretical and simulated symbol error rate (SER) curves are generated by varying the EsNo from 0 to 14 dB. Remember that if an imaginary component is added to 4-PAM, also with four levels, we get 16-QAM which is a complex modulation with four bits per symbol ($k = 4$) (Figs. 2.3 and 2.4).

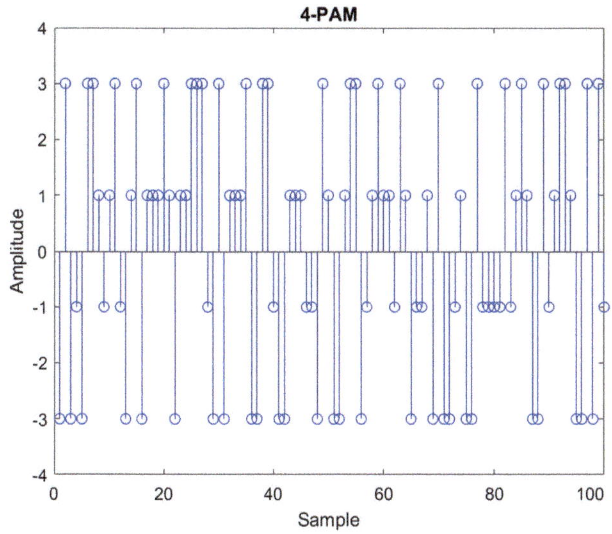

Fig. 2.3 Pulse amplitude modulation (with four levels)

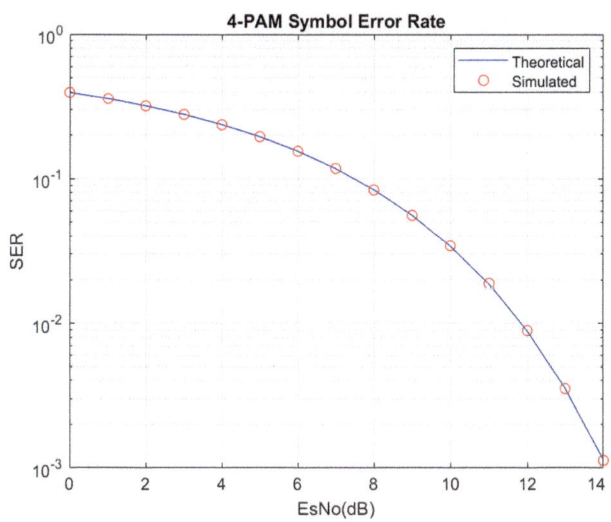

Fig. 2.4 Symbol error rate of 4-PAM

```
%%%%%%%%%%%%%%%%%%%%%%%%%%%%%%%%%%%%
%              PULSE AMPLITUDE MODULATION
%                     BER CALCULATION
%           l is the length of the symbol vector
%                  Es is the symbol energy
%         EsNo is the symbol energy to Noise PSD ratio
%                  Copyright 2020 RAYmaps
%%%%%%%%%%%%%%%%%%%%%%%%%%%%%%%%%%%%
clear all
close all

l=1e6;
Es=5;
EsNo_dB=10;
EsNo=10^(EsNo_dB/10);

% Symbol Generation for 4-PAM
s=rand(1,l);
for n=1:length(s)
  if s(n)<0.25
    s(n)=-3;
  elseif s(n)<0.50
    s(n)=-1;
  elseif s(n)<0.75
    s(n)=+1;
  else
    s(n)=+3;
  end
end

% Additive White Gaussian Noise
sigma=sqrt(Es/(2*EsNo));
r=s+sigma*randn(1,l);

% Symbol Detection for 4-PAM
for n=1:length(r)
  if r(n)<-2
    s_est(n)=-3;
  elseif r(n)<0
    s_est(n)=-1;
  elseif r(n)<2
    s_est(n)=+1;
  else
    s_est(n)=+3;
  end
end

% SER Calculation
simulatedSER=(l-sum(s==s_est))/l
theoreticalSER=0.75*erfc(sqrt(0.2*(EsNo)))
%%%%%%%%%%%%%%%%%%%%%%%%%%%%%%%%%%%%
```

Note: Please note that the following decision boundaries are used in 4-PAM.
-2 between -3 and -1
0 Between -1 and $+1$
$+2$ between $+1$ and $+3$

Optimized and Generalized Code for M-PAM SER

In the above code for calculation of symbol error rate (SER) for 4-PAM we have used a “for loop” as it is conceptually the simplest. But it is well known that the execution time of a “for loop” is quite high. It is desirable to replace the “for loop” with matrix manipulations. This is shown in the code below. In addition to simplification of code we have also converted it into a generalized form where any M-PAM ($M = 2, 4, 8, 16 \ldots$) can be simulated. It must also be noted that complex modulation schemes such as 64-QAM can also be implemented in a similar manner (symbols would have the form $a + jb$ where a and b would be generated independently as shown below). Also calculated is the theoretical SER for M-PAM which matches well with the simulation.

```
%%%%%%%%%%%%%%%%%%%%%%%%%%%%%%%%%%%%%%%%%%%%%%%%
%               PULSE AMPLITUDE MODULATION
%                        M-PAM
%               GENERALIZED IMPLEMENTATION
%               l is the length of the symbol vector
%                  Es is the symbol energy
%           EsNo is the symbol energy to Noise PSD ratio
%                  M is the number of levels
%                  Copyright 2020 RAYmaps
%%%%%%%%%%%%%%%%%%%%%%%%%%%%%%%%%%%%%%%%%%%%%%%%
clear all
close all

M=4;
levels=(1:M)*2-(M+1);
Es=sum((abs(levels)).^2)/length(levels);
EsNo_dB=10;
EsNo=10^(EsNo_dB/10);
l=1e6;

% Symbol Generation
s=2*floor(M*rand(1,l))-(M-1);

% Additive White Gaussian Noise
sigma=sqrt(Es/(2*EsNo));
r=s+sigma*randn(1,l);

% Symbol Detection
[A,B]=min([abs(levels'-r)]);
s_est=B*2-(M+1);

% SER Calculation
simulatedSER=(l-sum(s==s_est))/l
Ef=erfc(sqrt(EsNo/Es));
theoreticalSER=(2/M)*(1/2)*Ef+((M-2)/M)*(1)*Ef
%%%%%%%%%%%%%%%%%%%%%%%%%%%%%%%%%%%%%%%%%%%%%%%%
```

Note:

Note that in theoretical SER calculation for M-PAM the probability of symbol error at the two edges is half of the probability of symbol error at the (M-2) central positions. This is because the boundary symbols have only one neighbor, whereas the central ones have two. But we can use Ef = erfc(sqrt(EsNo/Es)) as a first estimate for symbol error probability, and this becomes successively more accurate as the constellation size (M) increases.

2.4 Minimum Shift Keying Bit Error Rate in AWGN

Minimum Shift Keying (MSK) is a type of continuous phase modulation (CPM) that has been used in many wireless communication systems. To be more precise it is Continuous Phase Frequency Shift Keying (CPFSK) with two frequencies f_1 and f_2. The frequency separation between the two tones is the minimum allowable while maintaining orthogonality and is equal to half the bit rate (or symbol rate, as both are the same). The frequency deviation is given as $\Delta f = R_b/4$. The two tones have frequencies of $f_c \pm \Delta f$, where f_c is the carrier frequency. MSK is sometimes also visualized as Offset QPSK (OQPSK) but we will not go into its details here.

As is known continuous phase modulations typically have a constant envelope. That is there is no Amplitude Modulation (AM), and the signal is not affected by distortion of carrier amplitude by fading or non-linear amplification. Continuity of phase also means that there are no abrupt changes in phase and sidelobe levels are very low. It must be noted however that in MSK the width of the main lobe is wider than in QPSK (but less than BPSK). The sidelobe level of MSK can be further reduced by using a Gaussian filer. The modulation so obtained is called Gaussian Minimum Shift Keying (GMSK).

GMSK modulation was adopted by GSM standard (a 2G standard) but has now gone out of favor as more spectrally efficient waveforms such as OFDM/QAM have taken over. The sidelobe level depends upon the symbol time (T) and bandwidth (B) product which is abbreviated as BT. Smaller the value of BT lower is the sidelobe level. However, lower value of BT means greater intersymbol interference (ISI) which requires an equalizer such as Decision Feedback Equalizer (DFE). In GSM systems BT was kept at 0.3 whereas in Bluetooth it was 0.5. In this post we consider simple MSK without any filtering. The channel is considered to be AWGN but fading can also be easily introduced (Figs. 2.5 and 2.6).

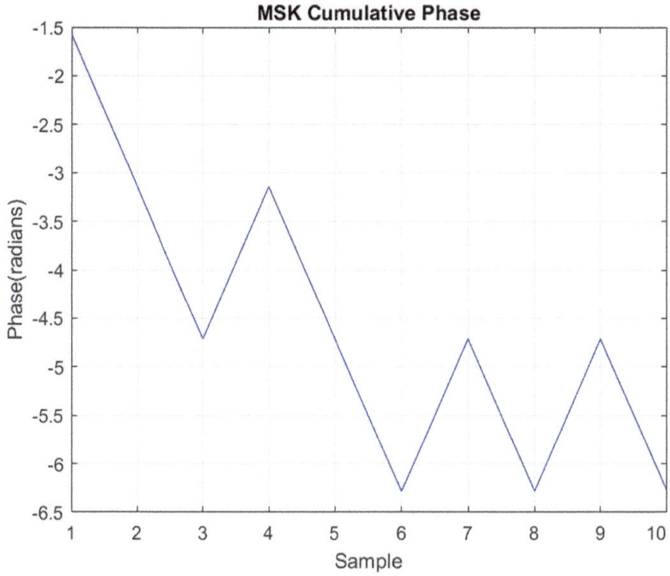

Fig. 2.5 MSK cumulative phase

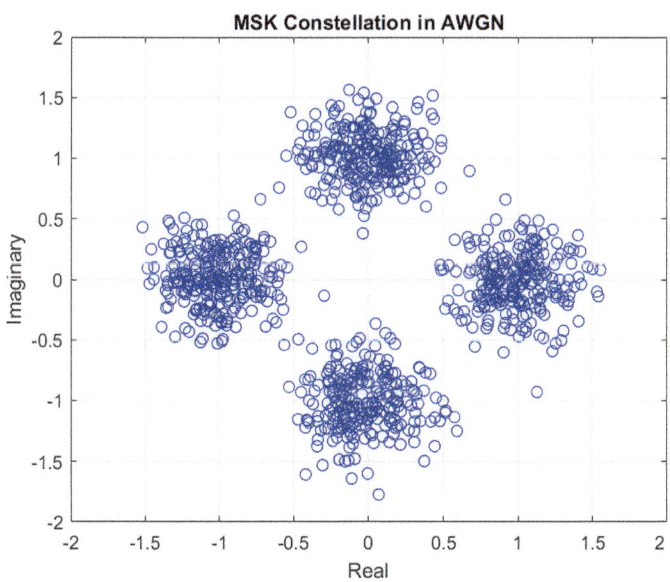

Fig. 2.6 MSK signal constellation in AWGN

MATLAB Code

```
%%%%%%%%%%%%%%%%%%%%%%%%%%%%%%%%%%%%%%%%%%%%%%%%%%
%                         MINIMUM SHIFT KEYING
%                          BER OF MSK IN AWGN
%                          Eb is the energy per bit
%                 EbNo is the energy per bit to noise PSD ratio
%                sigma is the standard deviation of AWGN noise
%                          Copyright 2020 RAYmaps
%%%%%%%%%%%%%%%%%%%%%%%%%%%%%%%%%%%%%%%%%%%%%%%%%%
clear all
close all

no_of_bits=5e6;
bits_in=round(rand(1,no_of_bits));
differential_bits=abs(diff([0, bits_in]));
bipolar_symbols=differential_bits*2-1;
cumulative_phase=(pi/2)*cumsum(bipolar_symbols);
tx_signal=exp(i*cumulative_phase);

Eb=1;
EbNodB=10;
EbNo=10^(EbNodB/10);
sigma=sqrt(Eb/(2*EbNo));
AWGN_noise=randn(1,no_of_bits)+i*randn(1,no_of_bits);
rx_signal=tx_signal+sigma*AWGN_noise;
rx_signal(2:2:end)=real(rx_signal(2:2:end));
rx_signal(1:2:end)=imag(rx_signal(1:2:end));

k=1:no_of_bits;
k=round(k/2);
multiplier=(-1).^(k+1);
demodulated_symbols=multiplier.*sign(rx_signal);
bits_out=demodulated_symbols*0.5+0.5;

ber_simulated=sum(bits_out!=bits_in)/no_of_bits
ber_theoretical=0.5*erfc(sqrt(EbNo))
%%%%%%%%%%%%%%%%%%%%%%%%%%%%%%%%%%%%%%%%%%%%%%%%%%
```

Simulation Results

Bit error rate (BER) simulation was carried out in MATLAB using one sample per symbol. This is done to keep the processing time to a minimum; however, the reader is encouraged to experiment with multiple samples per symbol. The environment under which the MSK BER is calculated is AWGN. Simulation for a Rayleigh fading channel can be done by adding the following two lines of code:

```
channel_coeff = (1/sqrt(2))*(randn(1,no_of_bits) + i*randn(1,no_of_bits));
rx_signal = abs(channel_coeff).*tx_signal + sigma*AWGN_noise;
```

Theoretical BER in Rayleigh fading environment can be calculated as:

ber_theoretical = 0.5*(1-sqrt(EbNo/(EbNo + 1))).

Theoretical and simulation results for both AWGN and Rayleigh environment match perfectly. It must be noted that BER results for both the environments exactly match with the results for BPSK. This is because both the modulations are essentially binary phase shift modulations, one using only real plane while other using both real and imaginary (alternatively) (Figs. 2.7 and 2.8).

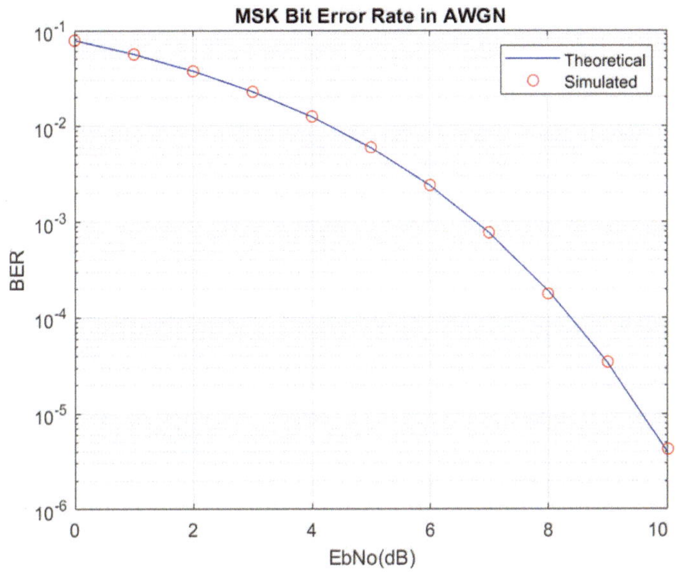

Fig. 2.7 MSK bit error rate in AWGN

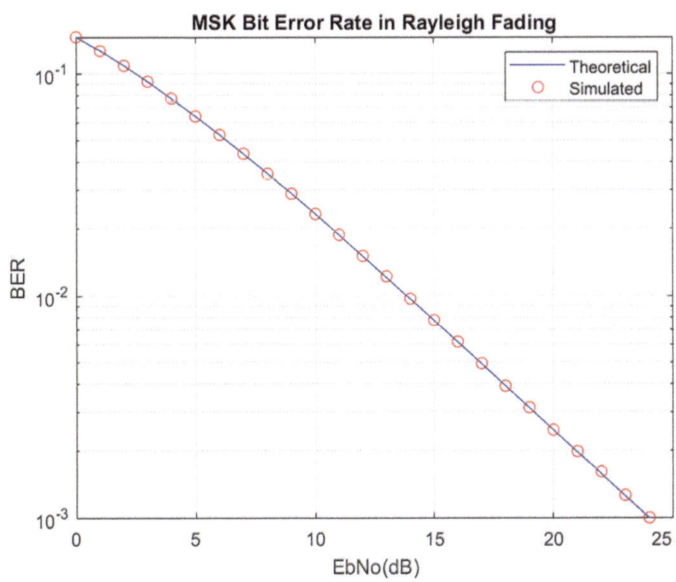

Fig. 2.8 MSK bit error rate in rayleigh fading

2.5 MSK Demodulation Using a Discriminator

The bit error rate (BER) performance of MSK modulation in AWGN and fading environments has been studied extensively, and various receiver architectures have been proposed [1]. It is widely believed that performance of non-coherent receivers is much worse than performance of coherent receivers in terms of BER. Although this is true to some extent but as we show in this section the difference in performance is not that much in case of Minimum Shift Keying (MSK). In fact, there is only a difference of about one dB in an AWGN environment at high signal to noise ratios (SNRs). The difference is somewhat larger in flat fading environment but given the simplicity of implementation of a non-coherent receiver the trade-off might be worth it.

Given below are the simulation results and MATLAB/Octave code for a discriminator-based MSK receiver architecture in an AWGN environment. It is seen that at low SNR the difference in performance is about 2 dB but this reduces to less than a dB in the high SNR region (there is a slight difference in how EbNo and SNR are defined but we use the terms interchangeably). The difference in performance in a flat fading environment is about 3–4 dB, keeping all other variables to be the same. It must also be noted that we are using one sample per symbol, the results change somewhat if we increase the number of samples per symbol. It is advisable that the reader studies the behavior for higher number of samples per symbol as this is desirable for a continuous phase modulation (CPM) that has smaller sidelobes as compared to other modulations such as BPSK/QPSK (Fig. 2.9).

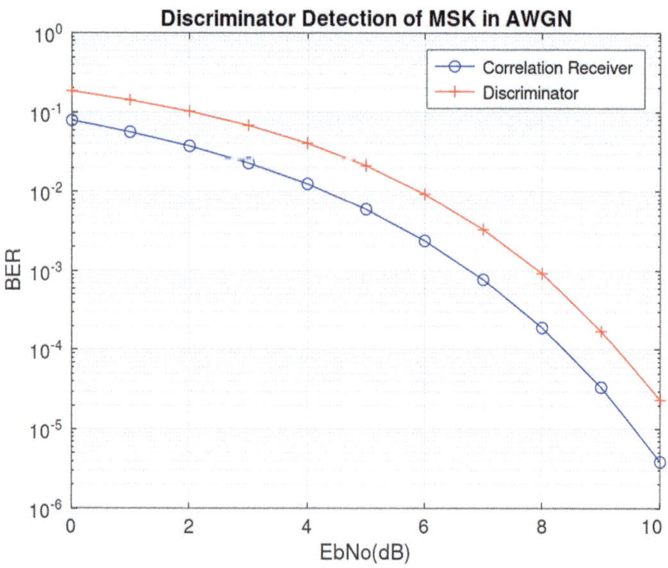

Fig. 2.9 MSK demodulation using a discriminator

```
%%%%%%%%%%%%%%%%%%%%%%%%%%%%%%%%%%%%%%%%%%%%%%%%%%%%%%%%%%%%
%                           BER OF MSK IN AWGN
%                      WITH DISCRIMINATOR DETECTOR
%                           Eb is the energy per bit
%               EbNodB is the energy per bit to noise PSD ratio in dBs
%      no_of_bits is the number of bits to be passed through the channel
%               sigma is the standard deviation of AWGN noise
%
%                           Copyright 2020 RAYmaps
%%%%%%%%%%%%%%%%%%%%%%%%%%%%%%%%%%%%%%%%%%%%%%%%%%%%%%%%%%%%
clear all
close all

no_of_bits=1e7;
Eb=1;
EbNodB=10;

% MODULATION
bits_in=round(rand(1,no_of_bits));
bipolar_symbols=bits_in*2-1;
commulative_phase=(pi/2)*cumsum(bipolar_symbols);
tx_signal=exp(i*commulative_phase);

% AWGN CHANNEL
EbNo=10^(EbNodB/10);
sigma=sqrt(Eb/(2*EbNo));
AWGN_noise=randn(1,no_of_bits)+i*randn(1,no_of_bits);
rx_signal=tx_signal+sigma*AWGN_noise;

% DEMODULATION
complex_rotation=([rx_signal, 1]).*conj([1, rx_signal]);
angle_increment=arg(complex_rotation(1:end-1));
detected_symbol=sign(angle_increment);
bits_out=detected_symbol*0.5+0.5;

% BER CALCULATION
ber_simulated=sum(bits_out!=bits_in)/no_of_bits;
ber_theoretical=0.5*erfc(sqrt(EbNo));
%%%%%%%%%%%%%%%%%%%%%%%%%%%%%%%%%%%%%%%%%%%%%%%%%%%%%%%%%%%%
```

Note:

1. Discriminator detector just outputs a $+1$ if the phase is increasing and outputs a -1 if the phase is decreasing.
2. In the code above we first find the advance of the exponential over the symbol period and then find the angle incremented. But discriminator can also be implemented by first finding the phase of the exponential and then taking a time derivative. Results remain exactly the same.
3. The results are much worse if we do oversampling, i.e., if we increase the number of samples per symbol. This is not totally understood at the moment and will be subject of a future discussion.
4. It must be noted that MSK phase advances by only $\pi/2$ degrees during a symbol period, and phase change is continuous. This means that there are no abrupt changes in phase, and the phase trajectory never crosses the origin.
5. One advantage of using non-coherent receiver architecture is that you do not require carrier phase synchronization.

2.6 Hamming Codes

We have previously discussed modulation and demodulation in wireless communications, now we turn our attention to channel coding. We know that in a wireless channel the transmitted information gets corrupted due to noise and fading, and we get what are called bit errors. One way to overcome this problem is to transmit the same information multiple times. In coding terminology this is called a repetition code. But this is not recommended as it results in reduced data rate and reduced spectral efficiency.

In this post we discuss Hamming (7,4) Code which transmits 4 information bits for every 7 bits transmitted, resulting in a code rate of 4/7. The 3 additional bits are called parity bits and these protect against single bit errors in the channel. This is called a systematic code since after performing the coding operation the information bits are preserved, parity bits are only appended to the information bits.

At the receiver we implement two decoding techniques, namely syndrome decoding and maximum likelihood decoding and compare the bit error rate with no coding case (BPSK modulation is assumed). In the first case syndrome is calculated at the receiver, which should be all zero if no error has occurred in the transmission. If a nonzero term appears in the syndrome, then it means that an error has occurred and a lookup table can be used to correct the error. It must be noted that only single bit errors can be corrected using this technique (here $d_{\min}$ is 3 and $t = (3–1)/2$).

The reason that this technique works is that the generator matrix at the transmitter is orthogonal to parity check matrix at the receiver, which is used in the calculation of syndrome. Next, we consider maximum likelihood decoding or soft decision decoding. This is a brute force method in which we search for the combination of symbols that have the minimum distance from the received symbols. This is done before the decision stage in the receiver as some information is lost in the decision stage.

The second method described above is based on Euclidean distance rather than Hamming distance. Euclidean distance is calculated between the possible transmitted symbols and the received symbols, whereas Hamming distance is calculated between the possible transmitted bits and received bits. As expected, maximum likelihood decoding performs much better than syndrome-based decoding which can detect only one bit error at a time. In fact, at low signal to noise ratio syndrome-based method is even worse than no coding case.

Note:

1. We have assumed BPSK modulation in the simulations but any other modulation format can be easily incorporated. In reality channel coding provides the leverage to go to higher modulation formats, resulting in higher spectral efficiency.
2. Single bit errors only need to be corrected for the four possible erroneous message sequences. Errors in parity bits can be ignored since they do not influence the bit error rate.
3. Hard decision decoding does not make full use of the information available; e.g., if we have BPSK modulation ($s = \pm 1$), there is no difference between a $+ 0.1$ and $+ 0.5$ after a bit decision is made. But soft decision decoding gives more weightage to $+ 0.5$ than $+ 0.1$ (Table 2.1; Fig. 2.10).

Table 2.1 Output bit pattern for all possible message sequences

S. No.	m0	m1	m2	m3	p0	p1	p2
0	0	0	0	0	0	0	0
1	0	0	0	1	0	1	1
2	0	0	1	0	1	1	0
3	0	0	1	1	1	0	1
4	0	1	0	0	1	1	1
5	0	1	0	1	1	0	0
6	0	1	1	0	0	0	1
7	0	1	1	1	0	1	0
8	1	0	0	0	1	0	1
9	1	0	0	1	1	1	0
10	1	0	1	0	0	1	1
11	1	0	1	1	0	0	0
12	1	1	0	0	0	1	0
13	1	1	0	1	0	0	1
14	1	1	1	0	1	0	0
15	1	1	1	1	1	1	1

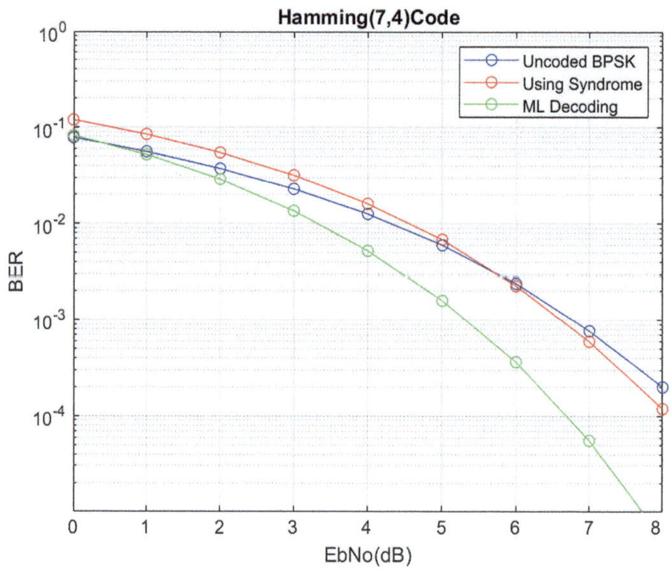

Fig. 2.10 Hamming (7,4) Code decoding using syndrome and ML decoding

```
%%%%%%%%%%%%%%%%%%%%%%%%%%%%%%%%%%%%%%%%%%%%%%%%%%%%%%%%%%%
%                ENCODING AND DECODING USING HAMMING CODE
%
%                          k is the number of message bits
%                          n is the number of encoded bits
%                              k/n is the code rate
%                           Copyright 2020 RAYmaps
%%%%%%%%%%%%%%%%%%%%%%%%%%%%%%%%%%%%%%%%%%%%%%%%%%%%%%%%%%%
clear all
close all

k=4;
n=7;
m=round(rand(1,k));

P=[1 0 1;
  1 1 1;
  1 1 0;
  0 1 1];
I=eye(k);
G=[I,P];
c=mod(m*G,2);
cx=2*c-1;

Eb=1.0*(n/k);
EbNo=10;
sigma=sqrt(Eb/(2*EbNo));
y=cx+sigma*randn(1,n);
d=y>0;
H=[P',eye(n-k)];
s=mod(d*H',2);


if s==([1 0 1])
 d(1)=not(d(1));
elseif s==([1 1 1])
 d(2)=not(d(2));
elseif s==([1 1 0])
 d(3)=not(d(3));
elseif s==([0 1 1])
 d(4)=not(d(4));

end
ber=sum(d(1:4)~=m)/4;
%%%%%%%%%%%%%%%%%%%%%%%%%%%%%%%%%%%%%%%%%%%%%%%%%%%%%%%%%%%
```

2.7 Convolutional Codes and Viterbi Decoding

In the previous post we discussed block codes and their decoding mechanisms. It was observed that with syndrome-based decoding there is only a minimal advantage over the no coding case. With Maximal Likelihood (ML) decoding there is significant improvement in performance but computational complexity increases exponentially with length of the code and alphabet size. This is where convolutional codes come to the rescue.

Convolutional codes when decoded using Viterbi algorithm (VA) provide significant gains over the no coding case. The performance is further improved by using soft metrics instead of hard metrics (Euclidean distance is used in this case instead of Hamming distance). In fact, the performance is even better than brute force ML decoded Hamming (7,4) Code discussed in the previous section.

The structure of the convolutional encoder used and state diagram is given below. The constraint length of this code is 3. This is the number of input bits that are used to generate the output bits at any instance of time. Higher the constraint length better is the performance but at the expense of computational complexity. The number of states of the convolutional code is given as 2^{K-1}, where K is the constraint length (here $K = 3$ and number of states is 4) (Fig. 2.11).

In this example 2 bits are generated at the output for 1 bit at the input resulting in a code rate of 1/2. This is catered for in the simulation by adjusting the signal to noise ratio so that there is no unfair advantage over the uncoded case. With soft

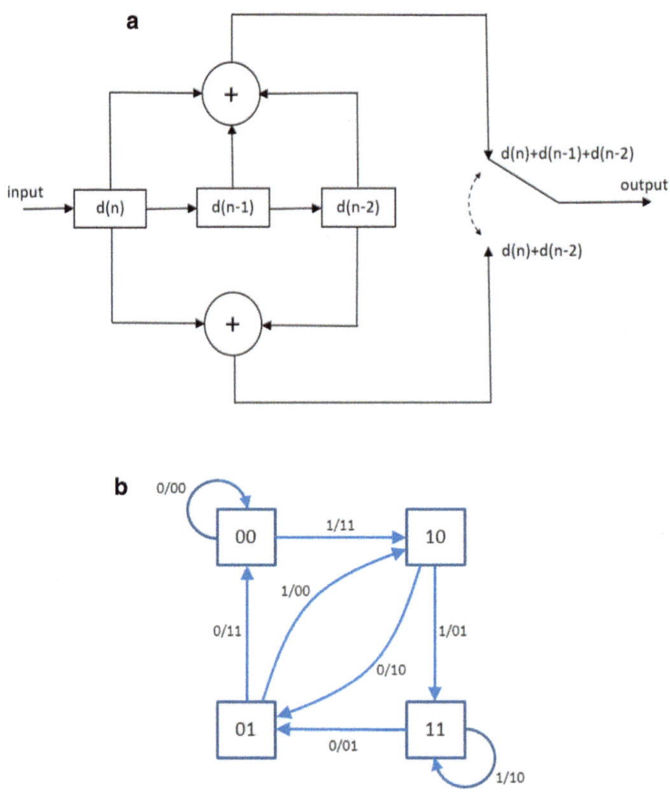

Fig. 2.11 Half rate convolutional encoder **a** Structure **b** State diagram

decoding a 2 dB performance improvement can be achieved over hard decoding at high signal to noise ratios. Just to be clear, in soft decoding, the distance metric is calculated before a decision is made on which symbol was most likely transmitted based on the observation.

Another important characteristic is the truncation length of the Viterbi algorithm. This decides how many state transitions are used in the decoding process. Higher the truncation length better is the performance of VA but after a certain point law of diminishing returns sets in. It must be noted that computational complexity of VA is lower than brute force ML decoding that is used for block codes but the performance is much better (at least for the case where soft decoding is used).

Note:

1. We have considered BPSK modulation in the simulation below but the code is general enough to accommodate any type of modulation. For complex modulations the Euclidean distance would be calculated in complex domain and AWGN noise added would also be complex.
2. It must be noted that for convolutional codes encoding process can be carried out by performing convolutional operation. But we have not used the MATLAB built-in function "conv" for this; rather, we simply perform modulo-2 addition. There is some advantage in using convolution operation, instead of modulo-2 addition, in terms of computation time.
3. In the block diagram above the rightmost two registers determine the state of the convolutional code. The leftmost register is the input. Together the three bits determine the output at each instant (Fig. 2.11).

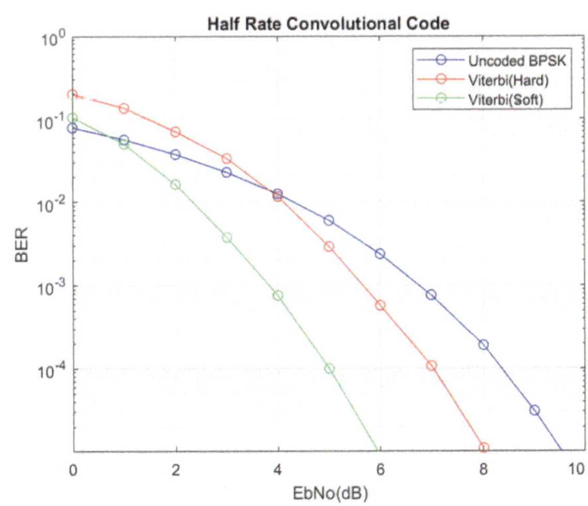

Fig. 2.12 Half rate convolutional code with viterbi decoding

```
%%%%%%%%%%%%%%%%%%%%%%%%%%%%%%%%%%%%%%%%%%%%%%%%%%%%%%%%%%%%%%%%%%%%%
%                ENCODING AND DECODING OF 1/2 CONVOLUTIONAL CODE
%
%                            Number of states is 4
%                   For each input bit there are 2 output bits
%                           Constraint length is 3
%                         Copyright 2020 RAYmaps
%%%%%%%%%%%%%%%%%%%%%%%%%%%%%%%%%%%%%%%%%%%%%%%%%%%%%%%%%%%%%%%%%%%%%
clear all
close all

% Encoding and Modulation
ip_length=50000;
msg_bits=round(rand(1,ip_length));
msg_bits=[0,0,msg_bits];
c=[];

for n=1:ip_length
  c1(n)=mod(msg_bits(n)+msg_bits(n+1)+msg_bits(n+2),2);
  c2(n)=mod(msg_bits(n)+msg_bits(n+2),2);
  c=[c,c1(n),c2(n)];
end
cx=2*c-1;

% Noise Addition and Demodulation
Eb=2.0;
EbNo=10;
sigma=sqrt(Eb/(2*EbNo));
y=cx+sigma*randn(1,2*ip_length);
d=y>0;

% Path Metric and Branch Metric Calculation
pm1(1)=0;
pm2(1)=1000;
pm3(1)=1000;
pm4(1)=1000;

d=[d,0,0,0,0];
for n=1:ip_length+2
  bm11=sum(abs([d(2*n-1),d(2*n)]-[0,0]));
  bm13=sum(abs([d(2*n-1),d(2*n)]-[1,1]));
  bm21=sum(abs([d(2*n-1),d(2*n)]-[1,1]));
  bm23=sum(abs([d(2*n-1),d(2*n)]-[0,0]));
  bm32=sum(abs([d(2*n-1),d(2*n)]-[1,0]));
  bm34=sum(abs([d(2*n-1),d(2*n)]-[0,1]));
  bm42=sum(abs([d(2*n-1),d(2*n)]-[0,1]));
  bm44=sum(abs([d(2*n-1),d(2*n)]-[1,0]));
```

```
    pm1_1=pm1(n)+bm11;
    pm1_2=pm2(n)+bm21;
    pm2_1=pm3(n)+bm32;
    pm2_2=pm4(n)+bm42;
    pm3_1=pm1(n)+bm13;
    pm3_2=pm2(n)+bm23;
    pm4_1=pm3(n)+bm34;
    pm4_2=pm4(n)+bm44;

    if pm1_1<pm1_2
       pm1(n+1)=pm1_1;
       tb_path(1,n)=0;
    else
       pm1(n+1)=pm1_2;
       tb_path(1,n)=1;
    end

    if pm2_1<pm2_2
       pm2(n+1)=pm2_1;
       tb_path(2,n)=0;
    else
       pm2(n+1)=pm2_2;
       tb_path(2,n)=1;
    end

    if pm3_1<pm3_2
       pm3(n+1)=pm3_1;
       tb_path(3,n)=0;
    else
       pm3(n+1)=pm3_2;
       tb_path(3,n)=1;
    end

    if pm4_1<pm4_2
       pm4(n+1)=pm4_1;
       tb_path(4,n)=0;
    else
       pm4(n+1)=pm4_2;
       tb_path(4,n)=1;
    end
end

[last_bm,last_state]=min([pm1(n+1),pm2(n+1),pm3(n+1),pm4(n+1)]);
m=last_state;
```

```
% Traceback and Decoding
for n=ip_length+2:-1:1
  if m==1
    if tb_path(m,n)==0
      op_bits(n)=0;
      m=1;
    elseif tb_path(m,n)==1
      op_bits(n)=0;
      m=2;
    end

  elseif m==2
    if tb_path(m,n)==0
      op_bits(n)=0;
      m=3;
    elseif tb_path(m,n)==1
      op_bits(n)=0;
      m=4;
    end

  elseif m==3
    if tb_path(m,n)==0
      op_bits(n)=1;
      m=1;
    elseif tb_path(m,n)==1
      op_bits(n)=1;
      m=2;
    end

  elseif m==4
    if tb_path(m,n)==0
      op_bits(n)=1;
      m=3;
    elseif tb_path(m,n)==1
      op_bits(n)=1;
      m=4;
    end
  end
end

ber=sum(msg_bits(3:end)~=op_bits(1:end-2))/ip_length
%%%%%%%%%%%%%%%%%%%%%%%%%%%%%%%%%%%%%%%%%%%%%%%%%%%%%%%%%%%%%%%%%%%%%%%%%%%%%%
```

bm: branch metric is the Hamming Distance/Euclidean Distance between the bits/symbols received and bits/symbols corresponding to a certain state transition.

pm: path metric is the total cost of a certain path using Hamming Distance or Euclidean Distance. The path metric for a state is the previous path metric plus the current branch metric that results in the lowest cost to reach this state.

Main Takeaway: A convolutional encoder introduces memory in the bit stream by allowing only certain state transitions and certain outputs at each step. Viterbi decoder makes use of this memory by estimating which sequence of state transitions was most likely to have been made based upon the observations. It breaks down a complex problem into smaller parts, each of one which is quite simple to solve. This is much easier than brute force search that we used in the case of Hamming codes.

2.8 Low-Density Parity Check Codes

We have previously discussed block codes and convolutional codes and their coding and decoding techniques particularly syndrome-based decoding and Viterbi decoding. Now we discuss an advanced form of block codes known as low-density parity check (LDPC) codes. These codes were first proposed by Robert Gallager in 1960 but they did not get immediate recognition as they were quite cumbersome to code and decode. But in 1995 the interest in these codes was revived, after discovery of turbo codes. Both these codes approach the Shannon Limit and have been adopted in many wireless communication systems including 5G (specifically LDPC codes and polar codes are used in 5G).

The name LDPC originates from the fact that the parity check matrix of LDPC codes is a sparse matrix with a few ones and lots of zeros, e.g., a size 1000 × 2000 parity check matrix might have 3 ones per column and 6 ones per row. This simplifies coding and decoding of LDPC codes. Another point to mention here is that LDPC codes draw their inspiration from turbo codes which are iteratively decoded, successively getting closer to the Shannon limit. LDPC codes can be decoded in a number of ways but the most common is serial decoding and parallel decoding, both iterative in nature [2]. We discuss parallel decoding in this post. If you understand coding and decoding of block codes, then half the problem is solved. Let's consider we have 6 data bits ($d_1, d_2, d_3, d_4, d_5, d_6$) and 5 parity check bits (p_1, p_2, p_3, p_4, p_5) are calculated as per the (Table 2.2).

The advantage of such a scheme is that each data bit is protected by multiple parity check bits so the possibility of an error going undetected is reduced. The equations that govern the calculation of parity bits are given below. Here the bit addition operations are performed modulo-2 (equivalent to XOR operation).

$$\begin{aligned} p_1 &= d_1 \oplus d_2 \oplus d_3 \\ p_2 &= d_4 \oplus d_5 \oplus d_6 \\ p_3 &= d_1 \oplus d_4 \\ p_4 &= d_2 \oplus d_5 \\ p_5 &= d_3 \oplus d_6 \end{aligned}$$

Table 2.2 Structure for calculating parity check bits

d_1	d_2	d_3	p_1
d_4	d_5	d_6	p_2
p_3	p_4	p_5	

It must be noted that during each transmission 6 data bits are transmitted as it is and 5 parity bits are appended to the data bits resulting in a block size of 11. As discussed previously such a code is called systematic code and can easily be encoded and decoded using a generator matrix $G = (I|C)$ and parity check matrix $H = (C^T|I)$, respectively. Another important metric of a code is the code rate which governs how much redundancy is added to the uncoded data bits. The code rate in the above example is $r = k/n = 6/11$. Lower the code rate higher is the redundancy and stronger is the code but this is at the expense of spectral efficiency. Typical code rates used in telecom systems are 1/3, 1/2, 2/3, etc.

Just as in block codes, error detection and correction can be performed by calculating the syndrome but BER performance using this technique is not that great. Maximum likelihood technique (remember brute search) can be used but it is very computationally complex for large block sizes. Here comes to our rescue a concept from probability theory called Bayes theorem. According to this theorem we can use posterior probabilities to decide which symbol was most likely to have been transmitted. After some simplifications to the Bayes theorem and some more mathematical manipulations we have the following equality (L is defined below).

$$L(x_1 \oplus x_2) = \text{sgn}[L(x_1)]\text{sgn}[L(x_2)] \min[|L(x_1)|, |L(x_2)|].$$

With this equality and parity check equations given above, the extrinsic information of each decoder can be calculated separately (decoder-I operates row-wise and decoder-II operates column-wise in the above table). This extrinsic information is then used to refine the estimation of the symbol that was most likely to have been transmitted. The sign is used to estimate the transmitted symbol sign and the absolute value provides the information about the reliability of the decoded symbol. As this process is iteratively performed the estimate of the transmitted symbol becomes more and more accurate (we have used up to 25 iterations in our simulations with improving results).

Let us explain the above concepts with an example. Let us consider the following parity check equation of encoder-I.

$$p_1 = d_1 \oplus d_2 \oplus d_3$$

If we want to estimate the value of d_1, we can rearrange the above equation as follows.

$$d_1 = p_1 \oplus d_2 \oplus d_3.$$

But in this equation, we have XOR operation being performed on the data bits d_2, d_3 and parity bit p_1. How do we perform the same operation when we have soft metrics? For this we use the following equality.

$$L(d_1) = L(p_1 \oplus d_2 \oplus d_3)$$
$$= \text{sgn}[L(p_1)]\text{sgn}[L(d_2)]\text{sgn}[L(d_3)] \min[|L(p_1)|, |L(d_2)|, |L(d_3)|].$$

The L values are described as LLR values or as Log Likelihood Ratios. If one knows the L values of p_1, d_2 and d_3, we can easily calculate the likelihood of d_1. The same procedure applies to all other bits, and it produces the required extrinsic information. If we use hard decisions in the above equations, we would get a much inferior bit error rate (this has been verified through simulation).

The BER results show that great performance gains can be achieved over the no coding case at low to moderate signal to noise ratios. When compared with other block codes it is seen that this scheme achieves the same BER performance as Hamming Code (7,4) with maximum likelihood decoding. This is quite an encouraging result because this is achieved with the simplest LDPC code we could imagine. With more complex codes the improvement in BER can be significant (Fig. 2.13). As a closing comment, I would like to say that Log Likelihood Ratio works by taking the logarithm of two probabilities. If the probability in the numerator (let's say p(x=+1)) is greater than the probability in the denominator (let's say p(x=–1)) i.e. the ratio is greater than one, we get a positive value at the output of the logarithm and vice versa.

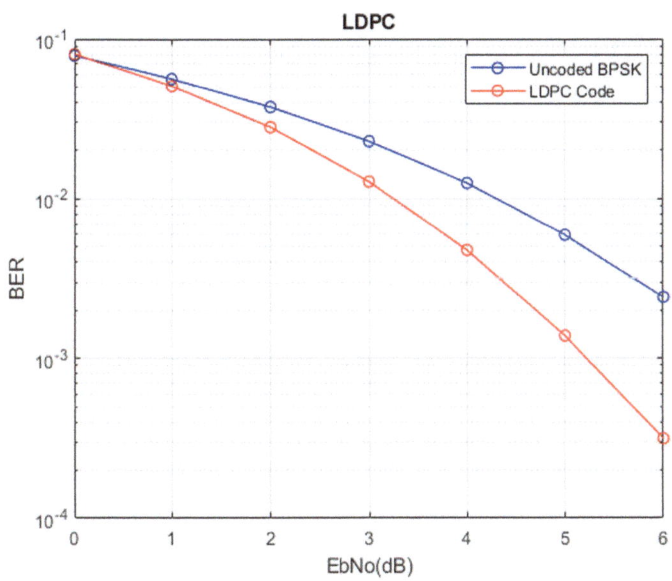

Fig. 2.13 BER performance of LDPC code compared with uncoded BPSK

```
%%%%%%%%%%%%%%%%%%%%%%%%%%%%%%%%%%%%%%%%%%%%%%%%%%%%%%
%                    ENCODING AND DECODING OF LDPC CODES
%
%                              k=6, n=11, r=6/11
%                           Copyright 2020 RAYmaps
%%%%%%%%%%%%%%%%%%%%%%%%%%%%%%%%%%%%%%%%%%%%%%%%%%%%%%
clear all
close all

k=6;
n=11;
Eb=1.0*(n/k);
EbNo=10;

a=round(rand(1,k));
C=[1 0 1 0 0;
  1 0 0 1 0;
  1 0 0 0 1;
  0 1 1 0 0;
  0 1 0 1 0;
  0 1 0 0 1];
I=eye(k,k);
G=[I,C];
b=mod(a*G,2);
c=1-2*b;
sigma=sqrt(Eb/(2*EbNo));
c=c+sigma*randn(1,n);

I=eye(n-k,n-k);
H=[C',I];
Lc_y=2/(sigma^2)*c;
L1=zeros(1,n);
L2=zeros(1,n);

for iter=1:25
 L1(1)=sign(c(2))*sign(c(3))*sign(c(7))*min(abs([c(2),c(3),c(7)]));
 L1(2)=sign(c(1))*sign(c(3))*sign(c(7))*min(abs([c(1),c(3),c(7)]));
 L1(3)=sign(c(1))*sign(c(2))*sign(c(7))*min(abs([c(1),c(2),c(7)]));
 L1(4)=sign(c(5))*sign(c(6))*sign(c(8))*min(abs([c(5),c(6),c(8)]));
 L1(5)=sign(c(4))*sign(c(6))*sign(c(8))*min(abs([c(4),c(6),c(8)]));
 L1(6)=sign(c(4))*sign(c(5))*sign(c(8))*min(abs([c(4),c(5),c(8)]));
 L1(7)=sign(c(1))*sign(c(2))*sign(c(3))*min(abs([c(1),c(2),c(3)]));
 L1(8)=sign(c(4))*sign(c(5))*sign(c(6))*min(abs([c(4),c(5),c(6)]));
 L1(9)=0;
 L1(10)=0;
 L1(11)=0;
```

```
L2(1)=sign(c(4))*sign(c(9))*min(abs([c(4),c(9)]));
L2(2)=sign(c(5))*sign(c(10))*min(abs([c(5),c(10)]));
L2(3)=sign(c(6))*sign(c(11))*min(abs([c(6),c(11)]));
L2(4)=sign(c(1))*sign(c(9))*min(abs([c(1),c(9)]));
L2(5)=sign(c(2))*sign(c(10))*min(abs([c(2),c(10)]));
L2(6)=sign(c(3))*sign(c(11))*min(abs([c(3),c(11)]));
L2(7)=0;
L2(8)=0;
L2(9)=sign(c(1))*sign(c(4))*min(abs([c(1),c(4)]));
L2(10)=sign(c(2))*sign(c(5))*min(abs([c(2),c(5)]));
L2(11)=sign(c(3))*sign(c(6))*min(abs([c(3),c(6)]));

L_sum=Lc_y+L1+L2;
c=L_sum;
Lc_y=2/(sigma^2)*c;

if sum(mod(H*((1-sign(c))/2)',2))==0
  break
end
end
c=sign(c);
d=(1-c)/2;
ber=sum(a(1:k)~=d(1:k))/k;
%%%%%%%%%%%%%%%%%%%%%%%%%%%%%%%%%%%%%%%%%%%%%%%%%%%%%%%%
```

Questions and Numerical Problems

1. Why do we need modulation, why can we not transmit zeros and ones directly? Give at least three reasons.
2. Why do we need channel coding, what are some of the pros and cons of channel coding?
3. What is the average energy per bit for a modulation scheme that transmits $\pm$ 1, what is this modulation scheme called?
4. What is the average energy per bit for a modulation scheme that transmits $\pm$ 1, $\pm$ 3, what is this modulation scheme called?
5. What are the advantages and disadvantages of using complex baseband equivalents in the simulation of communication systems?
6. What are the advantages and disadvantages of using a complex modulation instead of a real modulation?
7. What happens when the carriers of a complex modulation are not orthogonal? Explain in terms of the constellation diagram.
8. How can the bit error rate of a modulation scheme be improved? Explain in terms of the constellation diagram.
9. What are the advantages of continuous phase modulation?
10. Does the phase trajectory of the following modulation schemes pass through the origin: BPSK, QPSK, MSK?
11. What is that property of MSK due to which it can be demodulated using the Viterbi algorithm?
12. What is a repetition code? What are its advantages and disadvantages?

13. What is the difference between hard and soft decision decoding?
14. If the constraint length of a convolutional encoder is three, then how many states are there?
15. On what factors do the performance and computational complexity of Viterbi decoder depend?

Useful Links

1. Binary Phase Shift Keying Bit Error Rate in AWGN
https://www.raymaps.com/index.php/bit-error-rate-of-bpsk/
2. Pulse Amplitude Modulation Symbol Error Rate in AWGN
https://www.raymaps.com/index.php/pulse-amplitude-modulation-symbol-error-rate-in-awgn/
3. Minimum Shift Keying Bit Error Rate in AWGN
https://www.raymaps.com/index.php/minimum-shift-keying-bit-error-rate-in-awgn/
4. MSK Demodulation Using a Discriminator
https://www.raymaps.com/index.php/msk-demodulation-using-a-discriminator/
5. Hamming Codes
https://www.raymaps.com/index.php/hamming-codes/
6. Convolutional Codes and Viterbi Decoding
https://www.raymaps.com/index.php/convolutional-codes-and-viterbi-decoding/
7. Low-Density Parity Check Codes
https://www.raymaps.com/index.php/low-density-parity-check-codes/

References

1. Ahmed Y, Reed JH, Tranter WH, Michael Buehrer R (2003) A model based approach to demodulation of co-channel MSK signals. Globecom 2003, San Francisco, CA, USA.
2. Strutz T (2016) Low density parity check codes—an introduction. June 09.

Chapter 3
Diversity

3.1 Introduction

When a wireless signal propagates from the transmitter to the receiver its power falls off as squared of the distance; i.e., when the distance is varied from 1 to 100 m the power falls off by about 10,000 times or 40 dB. On top of it the signal undergoes fading which reduces its power even further. AWGN noise is added to this weak signal when it arrives at the receiver front end. Typical signal to noise ratio at the receiver is 0–20 dB, which may or may not be good enough to decode and demodulate the signal. Spatial diversity or simply diversity is a technique where multiple antennas are used to combine the signals at the receiver to improve the SNR. In modern wireless communications the concept has been further expanded to use multiple antennas at both the transmitter and receiver, referred to as multiinput multioutput (MIMO).

The basic concept of diversity is that if we have multiple independent channels of information, then the chances of all of the channels being in a bad state are much lower than the chance of a single channel being in a bad state; e.g., if the chances of a single channel being in a deep fade is 10%, then the chances of two channels being in a deep fade simultaneously is 1%. Taking this a step further, if there are four channels, then the chance of all of them being in a bad state is 0.01%. However, when using multiple receive antennas it must be ensured that there is sufficient spacing between them (at least half the wavelength) for the signals to be uncorrelated. Closer the antennas to each other, similar would be the signals received and lesser would be the diversity advantage.

Some of the diversity techniques employed are selection combining, threshold combining, equal gain combining, maximal ratio combining, etc. Selection combining is the simplest where the stronger of the two signals is selected, whereas in threshold combining the receiver switches to another signal when the currently selected signal drops below a predefined threshold. In equal gain combining, both the signals are added together giving them equal weighting. Maximal ratio combining, as the name suggests, gives more weight to the stronger signal and lesser weight to the

Y. Ahmed, *Recipes for Communication and Signal Processing*,
Signals and Communication Technology,
https://doi.org/10.1007/978-981-99-2917-7_3

weaker signal. Lastly we discuss a couple of techniques which employ multiple antennas at the transmitter instead of the receiver. This is particularly useful in scenarios where multiple antennas are placed at the base station and a single antenna is used at the mobile station.

It is seen that there is hardly any difference in bit error rate performance of equal gain and maximal ratio combining, with the latter being slightly better. Equal gain combining and transmit diversity (with the channel known at the transmit end) also have similar performance. When the channel is not known at the transmitter we can use the Alamouti scheme which is a full rate, complex, orthogonal Space Time Block Code (STBC) for two transmit antennas. However, the performance of Alamouti code is 3 dB worse than maximal ratio combining since it requires transmissions of the same symbol over two time slots and transmit energy has to be halved for each time slot to make a fair comparison. Search for efficient STBCs for more than two transmit antennas has been an active area of research since Alamouti code was first proposed in 1998.

3.2 Bit Error Rate of QPSK in AWGN

Simulating a QPSK system is equivalent to simulating two BPSK systems in parallel. So, there is no difference in bit error rate (BER). Since the simulation is at baseband, we multiply the in-phase and quadrature streams by 1 and j, respectively (instead of cos and sin carriers). At the receiver we just use the real and imag functions to separate the two symbol streams. The BER is the average BER of the two parallel streams (or we can just select one).

As in the case of BPSK we can show that the baseband representation (using 1 and j) is equivalent to using the passband representation (using cosine and sine carriers). Let's assume the following signal model for QPSK.

$$s(t) = a(t)\cos(2\pi f t) + b(t)\sin(2\pi f t),$$

where $a(t)$ and $b(t)$ contain the information to be transmitted over the channel. Now at the receiver we multiply this signal with cos to recover $a(t)$ and sin to recover $b(t)$. Multiplying by cosine carrier.

$$\begin{aligned}
\cos(2\pi f t)s(t) &= \cos(2\pi f t)[a(t)\cos(2\pi f t) + b(t)\sin(2\pi f t)] \\
&= a(t)\cos(2\pi f t)\cos(2\pi f t) + b(t)\sin(2\pi f t)\cos(2\pi f t) \\
&= \frac{a(t)}{2}[1 + \cos(4\pi f t)] + \frac{b(t)}{2}\sin(4\pi f t) \\
&= \frac{a(t)}{2} + \frac{a(t)}{2}\cos(4\pi f t) + \frac{b(t)}{2}\sin(4\pi f t).
\end{aligned}$$

After low-pass filtering (LPF) we get $a(t)/2$, which it the required in-phase component scaled by a constant 1/2. Similarly, we can find the quadrature component $b(t)$. Multiplying by sine carrier.

$$\begin{aligned}\sin(2\pi ft)s(t) &= \sin(2\pi ft)[a(t)\cos(2\pi ft) + b(t)\sin(2\pi ft)]\\ &= a(t)\sin(2\pi ft)\cos(2\pi ft) + b(t)\sin(2\pi ft)\sin(2\pi ft)\\ &= \frac{a(t)}{2}\sin(4\pi ft) + \frac{b(t)}{2}[1 - \cos(4\pi ft)]\\ &= \frac{a(t)}{2}\sin(4\pi ft) + \frac{b(t)}{2} - \frac{b(t)}{2}\cos(4\pi ft).\end{aligned}$$

Again, after low-pass filtering (LPF) we are left with the quadrature component scaled by the constant 1/2. So, we can conclude that the multiplication by the carrier terms at the transmitter and receiver is not required in simulation and we can simply transmit $a(t)$ and $b(t)$. We just have to make sure that $a(t)$ and $b(t)$ are orthogonal to each other so that they do not interfere with each other.

So, the transmitted QPSK signal would have the form.

$$s(t) = a(t) + jb(t).$$

The steps involved in the simulation are:

1. Generate a random sequence of symbols for the in-phase and quadrature components (-1 corresponding to binary value of 0 and $+1$ corresponding to binary value of 1). Add the in-phase and quadrature components in the form $a(t)+jb(t)$.
2. Generate complex samples of Additive White Gaussian Noise (AWGN) with the required variance (noise power = noise variance OR noise power = square of noise standard deviation OR noise power = noise power spectral density * signal bandwidth).
3. Add AWGN samples to the QPSK signal.
4. Detection is performed at the receiver by determining the sign of the in-phase and quadrature components.
5. And finally the bit error rate (BER) is calculated for the in-phase and quadrature components. Total bit error rate is the mean of the two values (Fig. 3.1).

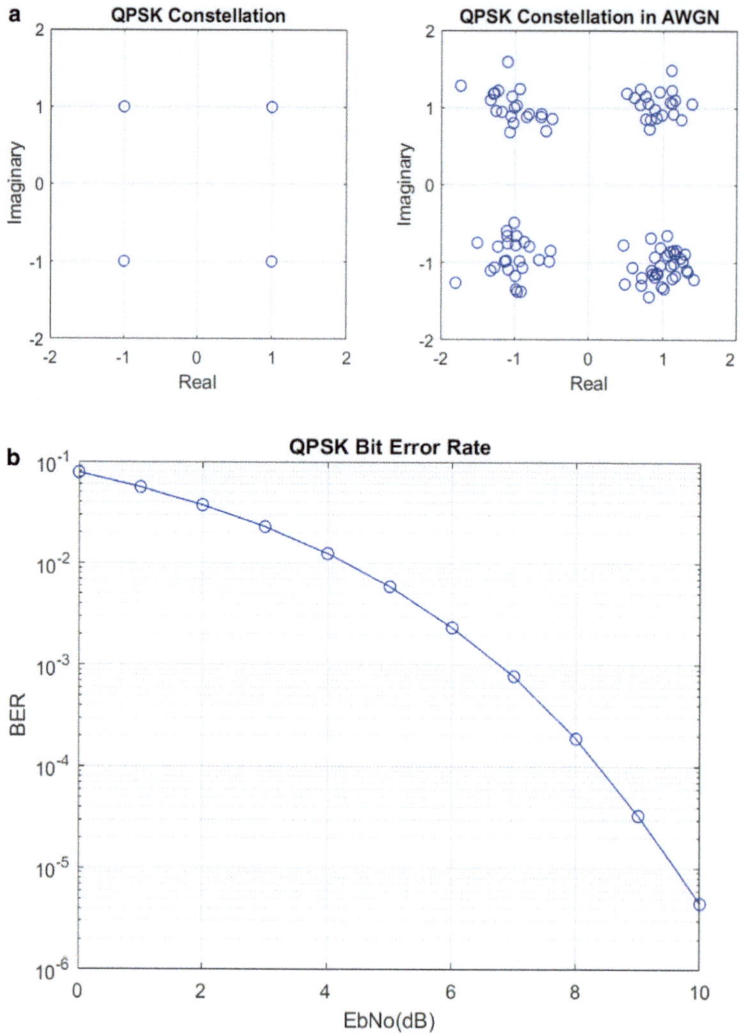

Fig. 3.1 **a** QPSK constellation **b** Bit error rate of QPSK as a function of EbNo

```
%%%%%%%%%%%%%%%%%%%%%%%%%%%%%%%%%%%%%%%%%%%%%%%%%%%%%%%%%%%%
%                    FUNCTION TO CALCULATE BER OF QPSK IN AWGN
%                         l is the length of the symbol sequence
%              EbNo is energy per bit to noise power spectral density ratio
%                          ber is the output bit error rate
%
%                              Copyright 2020 RAYmaps
%%%%%%%%%%%%%%%%%%%%%%%%%%%%%%%%%%%%%%%%%%%%%%%%%%%%%%%%%%%%
function[ber]=err_rate2(l,EbNo)
si=2*(round(rand(1,l))-0.5);
sq=2*(round(rand(1,l))-0.5);
s=si+j*sq;
n=(1/sqrt(2*10^(EbNo/10)))*(randn(1,l)+j*randn(1,l));
r=s+n;
si_=sign(real(r));
sq_=sign(imag(r));
ber1=(l-sum(si==si_))/l;
ber2=(l-sum(sq==sq_))/l;
ber=mean([ber1 ber2]);
return
%%%%%%%%%%%%%%%%%%%%%%%%%%%%%%%%%%%%%%%%%%%%%%%%%%%%%%%%%%%%
```

One final comment that I want to make is that bit error rate and symbol error rate is not always the same. Taking the example of QPSK a symbol error might occur when there is an error in the in-phase stream or the quadrature stream or both. So, it is not a one-to-one mapping!!!

Note:

1. For a QPSK constellation centered around the origin of the coordinate system, the decision boundaries are simply defined by the x and y axes.
2. The reason for the name QPSK is that there are four symbols in the constellation, each having one of four possible phases (45, 135, 225, 315).

3.3 Bit Error Rate of QPSK in Rayleigh Fading

So far, we have considered the bit error rate (BER) of BPSK and QPSK in an AWGN channel. Now we turn our attention to a Rayleigh fading channel which is a more realistic representation of a wireless communication channel. We consider a single tap Rayleigh fading channel which is good approximation of a flat fading channel, i.e., a channel that has flat frequency response (but varying with time). The complex channel coefficient is given as $(\alpha + j\beta)$ where α and β are Gaussian random variables with mean 0 and variance 0.5. We use the envelope of this channel coefficient in our simulation as any phase shift is easily removed by the receiver.

```
%%%%%%%%%%%%%%%%%%%%%%%%%%%%%%%%%%%%%%%%%%%%%%%%%%%%%%%%%%%%%
%                    FUNCTION TO CALCULATE BER OF QPSK IN FADING
%                         l is the length of the symbol sequence
%                EbNo is the energy per bit to noise power spectral density ratio
%                              ber is the output bit error rate
%
%                               Copyright 2020 RAYmaps
%%%%%%%%%%%%%%%%%%%%%%%%%%%%%%%%%%%%%%%%%%%%%%%%%%%%%%%%%%%%%
function[ber]=err_rate3(l,EbNo)
si=2*(round(rand(1,l))-0.5);
sq=2*(round(rand(1,l))-0.5);
s=si+j*sq;
n=(1/sqrt(2*10^(EbNo/10)))*(randn(1,l)+j*randn(1,l));
h=(1/sqrt(2))*((randn(1,l))+j*(randn(1,l)));
r=abs(h).*s+n;
si_=sign(real(r));
sq_=sign(imag(r));
ber1=(l-sum(si==si_))/l;
ber2=(l-sum(sq==sq_))/l;
ber=mean([ber1 ber2]);
return
%%%%%%%%%%%%%%%%%%%%%%%%%%%%%%%%%%%%%%%%%%%%%%%%%%%%%%%%%%%%%
```

It is observed that the BER for a Rayleigh fading channel is much higher than the BER for an AWGN channel. In fact, for Rayleigh fading, the BER curve is almost a straight line!!! (Fig. 3.2).

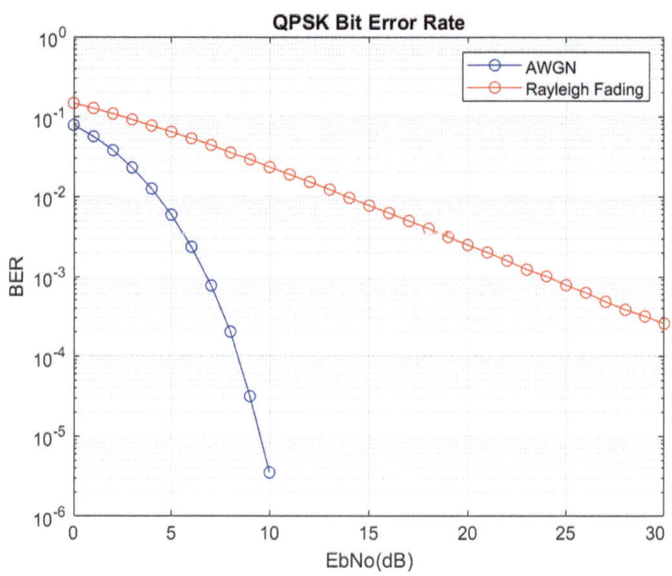

Fig. 3.2 Bit error rate of QPSK as a function of EbNo in Rayleigh fading

Note:

1. The input EbNo to the function is in dB so it is converted into linear scale by 10^(EbNo/10).
2. Noise is added in a Rayleigh fading channel as well. Noise is introduced by the receiver front end and is always present.

3.4 Equal Gain Combining in Rayleigh Fading

When wireless signals travel from a single transmit antenna to multiple receive antennas, they experience different fading conditions. While signal from one path may experience a deep fade the signal from another path may be stronger. Therefore, selecting the stronger of the two signals (selection combining, threshold combining) or adding the signals (equal gain combining, maximal ratio combining) would always yield much better results (lower bit error rate). However, there must be sufficient spacing between the different receive antennas for the received signals to be dissimilar (uncorrelated). In the simulation below we consider a 1-Tx, 2-Rx scenario. The signals arriving at the two receive antennas are added together before detection resulting in more than 10dB improvement in high SNR region (Fig. 3.3).

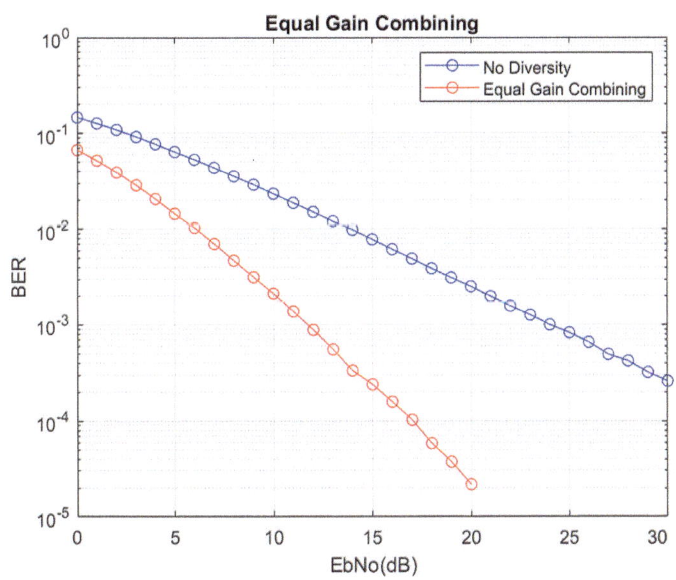

Fig. 3.3 Bit error rate of QSPK in Rayleigh fading using equal gain combining

```
%%%%%%%%%%%%%%%%%%%%%%%%%%%%%%%%%%%%%%%%%%%%%%%%%%%%%%%%%%%%%%%%%%%%%%%%%%%%%%%%%
%          FUNCTION TO CALCULATE BER OF QPSK USING EQUAL GAIN COMBINING
%
%                    l is the length of the symbol sequence
%          EbNo is the energy per bit to noise power spectral density ratio
%                        ber is the output bit error rate
%
%                          Copyright 2020 RAYmaps
%%%%%%%%%%%%%%%%%%%%%%%%%%%%%%%%%%%%%%%%%%%%%%%%%%%%%%%%%%%%%%%%%%%%%%%%%%%%%%%%%
function[ber]=err_rate4(l,EbNo)
si=2*(round(rand(1,l))-0.5);
sq=2*(round(rand(1,l))-0.5);
s=si+j*sq;
n1=(1/sqrt(2*10^(EbNo/10)))*(randn(1,l)+j*randn(1,l));
h1=(1/sqrt(2))*((randn(1,l))+j*(randn(1,l)));
n2=(1/sqrt(2*10^(EbNo/10)))*(randn(1,l)+j*randn(1,l));
h2=(1/sqrt(2))*((randn(1,l))+j*(randn(1,l)));
r1=abs(h1).*s+n1;
r2=abs(h2).*s+n2;
r=r1+r2;
si_=sign(real(r));
sq_=sign(imag(r));
ber1=(l-sum(si==si_))/l;
ber2=(l-sum(sq==sq_))/l;
ber=mean([ber1 ber2]);
return
%%%%%%%%%%%%%%%%%%%%%%%%%%%%%%%%%%%%%%%%%%%%%%%%%%%%%%%%%%%%%%%%%%%%%%%%%%%%%%%%%
```

Note:

1. Not only the signals on the two paths experience uncorrelated fading but the noise at the receiver front ends is also uncorrelated.
2. In reality the signals over both the paths would also experience random phase shifts but these can be removed before the combining process at the receiver.

3.5 Maximal Ratio Combining in Rayleigh Fading

We just saw the advantage an equal gain combiner (a combining scheme that just adds the signals after cophasing them) provides in a Rayleigh fading channel. Let's now look at a variant of this scheme called maximal ratio combining (MRC). In MRC the signals arriving at the receivers are weighted by the channel gains; i.e., a stronger signal is weighted more than a weaker signal before combining. It must be noted that in an actual system the received signals are both scaled and phase shifted thus an MRC receiver multiplies the received signals by the complex conjugate of the channel coefficients before addition.

```
%%%%%%%%%%%%%%%%%%%%%%%%%%%%%%%%%%%%%%%%%%%%%%%%%%%%%%%%%%%%%%%%%%
%           FUNCTION TO CALCULATE BER OF QPSK USING MAXIMAL RATIO COMBINING
%
%                       l is the length of the symbol sequence
%              EbNo is the energy per bit to noise power spectral density ratio
%                          ber is the output bit error rate
%
%                             Copyright 2020 RAYmaps
%%%%%%%%%%%%%%%%%%%%%%%%%%%%%%%%%%%%%%%%%%%%%%%%%%%%%%%%%%%%%%%%%%
function[ber]=err_rate4(l,EbNo)
si=2*(round(rand(1,l))-0.5);
sq=2*(round(rand(1,l))-0.5);
s=si+j*sq;
n1=(1/sqrt(2*10^(EbNo/10)))*(randn(1,l)+j*randn(1,l));
h1=(1/sqrt(2))*((randn(1,l))+j*(randn(1,l)));
n2=(1/sqrt(2*10^(EbNo/10)))*(randn(1,l)+j*randn(1,l));
h2=(1/sqrt(2))*((randn(1,l))+j*(randn(1,l)));
r1=abs(h1).*s+n1;
r2=abs(h2).*s+n2;
r=abs(h1).*r1+abs(h2).*r2;
si_=sign(real(r));
sq_=sign(imag(r));
ber1=(l-sum(si==si_))/l;
ber2=(l-sum(sq==sq_))/l;
ber=mean([ber1 ber2]);
return
%%%%%%%%%%%%%%%%%%%%%%%%%%%%%%%%%%%%%%%%%%%%%%%%%%%%%%%%%%%%%%%%%%
```

OR (using the complex channel coefficient)

```
%%%%%%%%%%%%%%%%%%%%%%%%%%%%%%%%%%%%%%%%%%%%%%%%%%%%%%%%%%%%%%%%%%
%           FUNCTION TO CALCULATE BER OF QPSK USING MAXIMAL RATIO COMBINING
%
%                       l is the length of the symbol sequence
%              EbNo is the energy per bit to noise power spectral density ratio
%                          ber is the output bit error rate
%
%                             Copyright 2020 RAYmaps
%%%%%%%%%%%%%%%%%%%%%%%%%%%%%%%%%%%%%%%%%%%%%%%%%%%%%%%%%%%%%%%%%%
function[ber]=err_rate4(l,EbNo)
si=2*(round(rand(1,l))-0.5);
sq=2*(round(rand(1,l))-0.5);
s=si+j*sq;
n1=(1/sqrt(2*10^(EbNo/10)))*(randn(1,l)+j*randn(1,l));
h1=(1/sqrt(2))*((randn(1,l))+j*(randn(1,l)));
n2=(1/sqrt(2*10^(EbNo/10)))*(randn(1,l)+j*randn(1,l));
h2=(1/sqrt(2))*((randn(1,l))+j*(randn(1,l)));
r1=h1.*s+n1;
r2=h2.*s+n2;
r=conj(h1).*r1+conj(h2).*r2;
si_=sign(real(r));
sq_=sign(imag(r));
ber1=(l-sum(si==si_))/l;
ber2=(l-sum(sq==sq_))/l;
ber=mean([ber1 ber2]);
return
%%%%%%%%%%%%%%%%%%%%%%%%%%%%%%%%%%%%%%%%%%%%%%%%%%%%%%%%%%%%%%%%%%
```

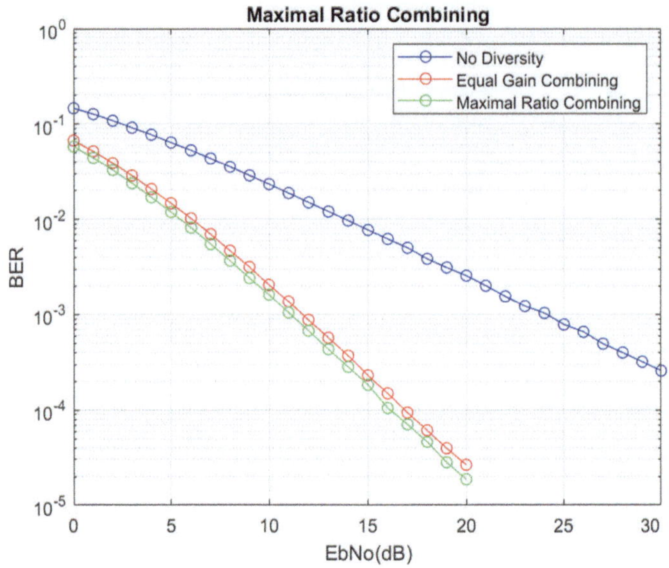

Fig. 3.4 Bit error rate of QPSK in Rayleigh fading using maximal ratio combining

We see that there is an incremental improvement in BER using MRC instead of EGC (1 dB can sometimes be significant) (Fig. 3.4).

Note:

1. The performance of MRC is the same using both the techniques given above.
2. Phase rotation of the noise component does not affect the BER performance.

3.6 Transmit Diversity Using Channel State Information

We saw that equal gain combining and maximal ratio combining result in tremendous improvement in bit error rate performance in a Rayleigh fading channel. These are received diversity schemes, i.e., schemes that work with multiple receive antennas. Now let us turn our attention to schemes that work with multiple transmit antennas. We know that the main aim of a combining scheme is to coherently add the signals. If the same signal is transmitted from multiple transmit antennas, the resulting signals would not add up coherently when they arrive at the receiver (remember that each path introduces a random phase shift). One solution to this problem is that the channel state information (CSI) be fed back to the transmitter. So, if this is done quickly enough, before the channel state changes, the phase of the signals at the transmit side could be pre-adjusted so that when these signals arrive at the receiver they combine constructively.

```
%%%%%%%%%%%%%%%%%%%%%%%%%%%%%%%%%%%%%%%%%%%%%%%%%%%%%%%%%%%%%%%%%%%%%%
%          FUNCTION TO CALCULATE BER OF QPSK USING CSI AT THE TRANSMITTER
%
%                      l is the length of the symbol sequence
%          EbNo is the energy per bit to noise power spectral density ratio
%                      ber is the output bit error rate
%
%                      Copyright 2020 RAYmaps
%%%%%%%%%%%%%%%%%%%%%%%%%%%%%%%%%%%%%%%%%%%%%%%%%%%%%%%%%%%%%%%%%%%%%%
function[ber]=err_rate5(l,EbNo)
si=2*(round(rand(1,l))-0.5);
sq=2*(round(rand(1,l))-0.5);
s=si+j*sq;
n=(1/sqrt(2*10^(EbNo/10)))*(randn(1,l)+j*randn(1,l));
h1=(1/sqrt(2))*((randn(1,l))+j*(randn(1,l)));
h2=(1/sqrt(2))*((randn(1,l))+j*(randn(1,l)));
sr1=(1/sqrt(2))*s.*(conj(h1)./abs(h1));
sr2=(1/sqrt(2))*s.*(conj(h2)./abs(h2));
r=h1.*sr1+h2.*sr2+n;
si_=sign(real(r));
sq_=sign(imag(r));
ber1=(l-sum(si==si_))/l;
ber2=(l-sum(sq==sq_))/l;
ber=mean([ber1 ber2]);
return
%%%%%%%%%%%%%%%%%%%%%%%%%%%%%%%%%%%%%%%%%%%%%%%%%%%%%%%%%%%%%%%%%%%%%%
```

It is observed that the above scheme has exactly the same bit error rate performance as equal gain combining. The reason for this is that in the above scheme the noise at the receiver is halved (single receiver means single noise component) but the transmit power is also halved from each of the transmit antennas (to keep the total transmit power same). Thus, it does not matter whether the phase adjustment happens at the receiver or the transmitter. But the important question is that can the channel state information be fed back to the transmitter quickly enough?

3.7 Alamouti Scheme

So, we have seen that multiple transmit antennas provide the same gain as multiple receive antennas if the channel state information can be fed back to the transmitter. But what if the channel state information cannot be fed back to the transmitter (or it can be done but not quickly enough). The solution to this problem is the so-called Alamouti Scheme. In this scheme two symbols are simultaneously transmitted from two transmit antennas and in the next time slot phase shifted versions of these two symbols are transmitted over the two transmit antennas. The channel is assumed to be quasi-static; i.e., it is static over the duration of two time slots but then changes for the next two time slots. A combining scheme is used at the receiver which separates the two symbols.

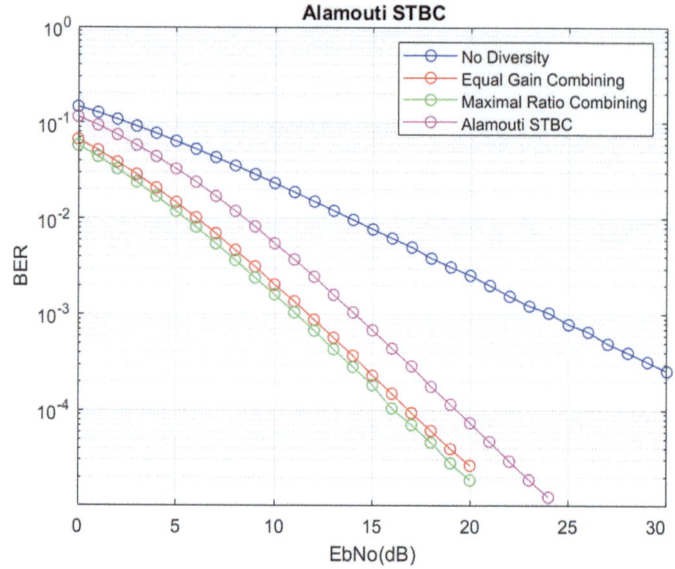

Fig. 3.5 BER of Alamouti scheme compared with no diversity, EGC and MRC schemes

```
%%%%%%%%%%%%%%%%%%%%%%%%%%%%%%%%%%%%%%%%%%%%%%%%%%%%%%%%%%%%%%
%                ALAMOUTI TRANSMIT DIVERSITY SCHEME
%                  l is the length of symbol sequence
%          EbNo is energy per bit to noise power spectral density ratio
%                    ber is output bit error rate
%                    Copyright 2020 RAYmaps
%%%%%%%%%%%%%%%%%%%%%%%%%%%%%%%%%%%%%%%%%%%%%%%%%%%%%%%%%%%%%%
function[ber]=err_rate7(l,EbNo)
s1i=2*(round(rand(1,l))-0.5);
s1q=2*(round(rand(1,l))-0.5);
s1=s1i+j*s1q;
s2i=2*(round(rand(1,l))-0.5);
s2q=2*(round(rand(1,l))-0.5);
s2=s2i+j*s2q;

n1=(1/sqrt(2*10^(EbNo/10)))*(randn(1,l)+j*randn(1,l));
n2=(1/sqrt(2*10^(EbNo/10)))*(randn(1,l)+j*randn(1,l));
h1=(1/sqrt(2))*((randn(1,l))+j*(randn(1,l)));
h2=(1/sqrt(2))*((randn(1,l))+j*(randn(1,l)));

r1=h1.*(sqrt(1/2)*s1)+h2.*(sqrt(1/2)*s2)+n1;
r2=-h1.*conj(sqrt(1/2)*s2)+h2.*conj(sqrt(1/2)*s1)+n2;
s1_=conj(h1).*r1+h2.*conj(r2);
s2_=conj(h2).*r1-h1.*conj(r2);
s1i_=sign(real(s1_));
s1q_=sign(imag(s1_));

ber1=(l-sum(s1i==s1i_))/l;
ber2=(l-sum(s1q==s1q_))/l;
ber=mean([ber1 ber2]);
return
%%%%%%%%%%%%%%%%%%%%%%%%%%%%%%%%%%%%%%%%%%%%%%%%%%%%%%%%%%%%%%
```

It is observed that this scheme is 3 dB worse (Fig. 3.5) than MRC (and transmit diversity with CSI). This reason for this is that unlike MRC the signals are transmitted from two transmit antennas, and thus the power is halved at each transmit antenna (this scheme is also approximately 3 dB worse than transmit diversity with CSI at transmitter because although both schemes transmit half the power from each source but in this scheme the noise power is doubled due to the combining scheme working over two time slots).

Questions and Numerical Problems

1. Derive the mathematical relationship for the probability of bit error for BPSK as a function of energy per bit to noise power spectral density ratio in AWGN environment.
2. Is the probability of bit error for BPSK the same as probability of bit error for QPSK in AWGN? Explain.
3. Is the probability of symbol error for BPSK the same as probability of symbol error for QPSK in AWGN? Explain.
4. Derive the mathematical relationship for the probability of bit error for BPSK as a function of energy per bit to noise power spectral density ratio in a Rayleigh fading environment.
5. Same information is passed from the transmitter to the receiver using two identical but independent channels. If the chance of failure on each channel is 10%, what is the chance of failure on both the channels simultaneously?
6. Same information is passed from the transmitter to the receiver using four identical but independent channels. If the chance of failure on each channel is 10%, what is the chance of failure on all of the channels simultaneously?
7. We briefly mentioned selection combining and threshold combining in the discussion on diversity but did not give the code. Simulate both these combining schemes for a range of signal to noise ratios and compare with the other schemes we have discussed.
8. Discuss the computational complexity of each of the diversity schemes discussed. Which works the best and which is the worst in terms of computational complexity and performance?
9. If there is a fast fading Rayleigh channel, would receive diversity still work, explain with an example.
10. If there is a fast fading Rayleigh channel, would transmit diversity still work, explain with an example.
11. Simulate the performance of at least two diversity schemes with imperfect channel state information (CSI). Comment on the performance.
12. Simulate the bit error rate performance of Alamouti scheme when the channel is not quasi-static. Comment on the performance.
13. Why is Alamouti scheme 3 dB worse than MRC combining? Why is Alamouti scheme 3 dB worse than transmit diversity with perfect CSI at the transmitter?
14. Why does a full rate, complex, orthogonal Space Time Block Code (STBC) only exist for two transmit antennas?
15. Simulate the performance of an STBC for four transmit antennas. You can compromise on the full rate of the code.

Useful Links

1. Bit Error Rate of QPSK in AWGN
   https://www.raymaps.com/index.php/bit-error-rate-of-qpsk/
2. Bit Error Rate of QPSK in Rayleigh Fading
   https://www.raymaps.com/index.php/bit-error-rate-of-qpsk-in-rayleigh-fading/
3. Equal Gain Combining in Rayleigh Fading
   https://www.raymaps.com/index.php/equal-gain-combining-in-rayleigh-fading/
4. Maximal Ratio Combining in Rayleigh Fading
   https://www.raymaps.com/index.php/maximal-ratio-combining-in-rayleigh-fading/
5. Transmit Diversity using Channel State Information
   https://www.raymaps.com/index.php/transmit-diversity-using-channel-state-information/
6. Alamouti Scheme
   https://www.raymaps.com/index.php/alamouti-scheme/

References

1. Rappaport, T. S.: Wireless Communications: Principles and Practice, Second Edition, Prentice Hall (2002)
2. Alamouti S.M.: A Simple Transmit Diversity Technique for Wireless Communications. IEEE Journal on Selected Areas in Communications. **16**(8): 1451–1458 (1998)

Chapter 4
Multicarrier

4.1 Introduction

A little bit of history is needed to set the stage for multicarrier.

As wireless standards have progressed from 1 to 5G, bandwidths have increased and voice quality has improved. Capacity of wireless systems in terms of data rate and number of users has also increased. This is true for all wireless standards except AMPS (1G) which was a pure analog system with 30 kHz bandwidth and voice-only transmission. GSM (2G) which was the first digital system allowed for voice and SMS to be sent. Both these standards, AMPS and GSM, used Frequency Division Multiple Access; i.e., users were differentiated based on the carrier frequency. But there is a small caveat, a 200 kHz GSM channel was divided into eight time slots, so a maximum of eight users could be accommodated at any time in one channel. GPRS, a 2.5 G standard, allowed for multiple time slots to be used by a single user and allowed for data transmission of 56 to 114 kbps.

At the time that Europe was happily adopting GSM technology and reaping the benefits of a digital system, a revolution was happening in the USA. For the first time Code Division Multiple Access or CDMA was being adopted by the cellular industry, a pure digital system, which used codes instead of frequencies to separate the users. This was still a 2G standard and was known as Interim Standard 95 (IS-95) or by its proprietary name cdmaOne. The man behind this new technology was Andrew J. Viterbi, who along with Irwin Jacobs, founded Qualcomm and who owned the key patents for CDMA. The biggest advantage of CDMA was that it did not have a fixed capacity as FDMA had and users were accommodated as they entered the system with a small hit to the quality of the signal. IS-95 achieved a data rate of up to 115 kbps over a 1.25 MHz bandwidth.

This shift from FDMA to CDMA continued and Europe engineered its own 3G standard by the name of Wideband CDMA or WCDMA. The main reason that Europe invented its own CDMA standard was that this way it did not have to pay royalties to Qualcomm which had a 3G standard by the name of CDMA2000. WCMDA and

Y. Ahmed, *Recipes for Communication and Signal Processing*,
Signals and Communication Technology,
https://doi.org/10.1007/978-981-99-2917-7_4

CDMA2000 provided much higher data rates and spectral efficiencies as higher bandwidths were available and higher-order modulation and coding schemes were used. Initial versions of WCDMA could achieve a data rate of 384 kbps, whereas CDMA2000 1× was able to achieve 154 kbps, a somewhat lower gain over its predecessor. Even higher data rates were possible by adopting 3.5/3.75 transitional technologies such as HSUPA/HSDPA and CDMA2000 1xEV-DO.

After the CDMA revolution of the 90s, another revolution was happening by the time 4G standards were being drafted. It consisted of two parts: Orthogonal Frequency Division Multiplexing (OFDM) as the modulation technique and advanced MIMO as the antenna technology. OFDM allowed for a wideband channel to be divided into narrow bins so the effect of frequency selective fading could be overcome. Advanced MIMO technologies allowed for more resistance to channel fading and higher multiplexing gain. This resulted in a spectral efficiency of 5 bps/Hz or even higher. Just to give you an example, with this spectral efficiency and a bandwidth of 20 MHz a data rate of 100 Mbps was easily achievable. WiMAX, another MIMO-OFDM standard that was the forerunner of LTE was proposed by the IEEE working group IEEE 802.16. WiMAX did not gain much popularity and is now used by a limited number of countries in Africa and Asia.

$$\text{FDMA} > \text{TDMA} > \text{CDMA} > \text{OFDMA}$$

This brings us to the current state of the wireless world that has seen wide deployment of 5G, and carriers and vendors are in a race to get there first. Two fundamental blocks of 5G, that would enable it to achieve data rates of 1 Gbps and higher, are massive MIMO and millimeter wave. Massive MIMO envisions that tens to hundreds of antennas would be used at the transmitter and receiver to achieve high spatial multiplexing. Millimeter wave bands that have bandwidths in excess of 1 GHz may also be utilized to increase the data rate. Experiments have shown that spectral efficiencies of 50 bps/Hz can be achieved with such configurations resulting in tens of gigabits of download speeds. But let's not forget that OFDM/OFDMA (OFMDA is simply OFDM used as a multiple access scheme) continues to be a key building block beyond 4G. Utilization of such high bandwidths, such as in millimeter wave, would only be possible if we have a solution to frequency selective fading. Multicarrier would be essential if we want to fully utilize the power of millimeter wave and massive MIMO.

4.2 BER of 64-QAM OFDM in AWGN

OFDM modulation works on the principle of converting a serial symbol stream to a parallel symbol stream with each symbol from the parallel set modulating a separate carrier. The spacing between the carriers is 1/T where T is the duration of the OFDM symbols (without cyclic prefix). This guarantees orthogonality of the carriers; i.e., there is no interference between carriers. The addition of orthogonal carriers modulated by parallel symbol streams is equivalent to taking the IFFT of the parallel symbol set. At the receiver the inverse operation of FFT is performed and then the parallel symbol stream is converted to a serial symbol stream.

The main advantage of this scheme is that one carrier (or set of carriers) may undergo severe fading but other carriers would be able to carry data. Equalization on these narrowband channels is also much easier than equalization of one wideband channel. Intersymbol interference (ISI) which affects the signal in the time domain is removed by adding a guard period between symbols, called cyclic prefix (CP). The CP converts linear convolution, which a fading channel performs, to circular convolution. This reduces the process of equalization at the receiver to simply a division operation (Table 4.1; Fig. 4.1).

1. **Random Data Generator:** A source of binary sequence that models a real data source such as a mobile, laptop and computer.
2. **Modulator:** Converts bits to symbols or in other words converts a sequence of ones and zeros to symbols of the form $a + jb$.
3. **Serial to Parallel:** Converts the output into a form suitable for IFFT operation. In the above example the length of the IFFT is 128.
4. **IFFT:** Inverse Fast Fourier Transform to convert frequency domain signal to time domain.
5. **Parallel to Serial:** Converts the output of the IFFT into serial form required for transmission.
6. **Add Cyclic Prefix:** Cyclic prefix converts linear convolution, which a fading channel performs, into circular convolution. This block is not implemented in the code below as no fading is assumed. This will be discussed in the next post.
7. **Channel:** Channel is assumed to be Additive White Gaussian Noise (AWGN) channel. No fading is implemented. Fading will be implemented in the next post.
8. **Remove Cyclic Prefix:** Remove the cyclic prefix that was added at the transmit end. This block is also not implemented here. This will be discussed in the next post.
9. **Serial to Parallel:** Just as above, this is needed to perform the FFT operation.

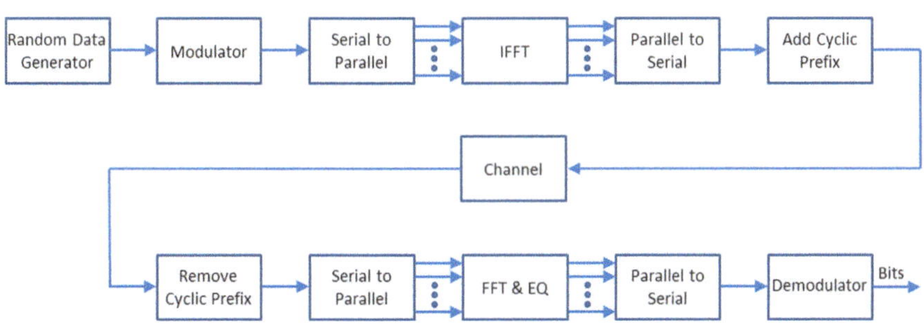

Fig. 4.1 OFDM modulator and demodulator block diagram

Table 4.1 LTE physical layer parameters

Transmission BW		1.25 MHz	2.5 MHz	5 MHz	10 MHz	15 MHz	20 MHz
Sub-frame duration		0.5 ms					
Sub-carrier spacing		15 kHz					
Sampling frequency		192 MHz (1/2 × 3.84 MHz)	3.84 MHz	7.68 MHz (2 × 3.84 MHz)	15 36 MHz (4 × 3.84 MHz)	23 04 MHz (6 × 3.84 MHz)	30.72 MHz (8 × 3.84 MHz)
FFT size		128	256	512	1024	1536	2048
OFDM sym per slot (short/long CP)		7/6					
CP length (usee/ samples)	Short	(4,69/9) × 6, (5.21/10) × 1	(4.69/18) × 6, (5.21/20) × 1	(4.69/36) × 6, (5.21/40) × 1	(4.69/72) × 6, (5.21/60) × 1	(4.69/108) × 6, (5.21/120) × 1	(4.69/144) × 6, (5.21/160) × 1
	Long	(16.67/32)	(16.67/64)	(1667/128)	(1667/256)	(1667/384)	(16 67/512)

10. **FFT and EQ:** Fast Fourier Transform to convert time domain signal to frequency domain. Equalization is not required as there is no fading. Equalization will be discussed in the next post.
11. **Parallel to Serial:** Converts the output of the FFT into a serial form required for demodulation.
12. **Demodulator:** Converts symbols into bits.

64-QAM is an important component of the LTE Air Interface (and now 5G as well) that promises higher data rates and spectral efficiencies. Combined with OFDM and MIMO it successfully combats the detrimental effects of the wireless channels and provides data rates in excess of 100 Mbps (peak data rate). Here, we discuss a simple example of 64-QAM modulation with OFDM in an AWGN channel. We assume a bandwidth of 1.25MHz which corresponds to an FFT size of 128.

As discussed previously, with proper normalization of IFFT and FFT operations the performance of OFDM in AWGN is the same as the performance of the underlying modulation scheme (64-QAM in this case) (Fig. 4.2). We have not even introduced the cyclic prefix in our simulation because without a fading channel there is no ISI and cyclic prefix (CP) is of no use. We will introduce the CP when we turn our attention to fading channels.

It must be noted that although IFFT and FFT are linear inverses of each other, proper normalization is required to maintain the signal levels at the transmitter and receiver.

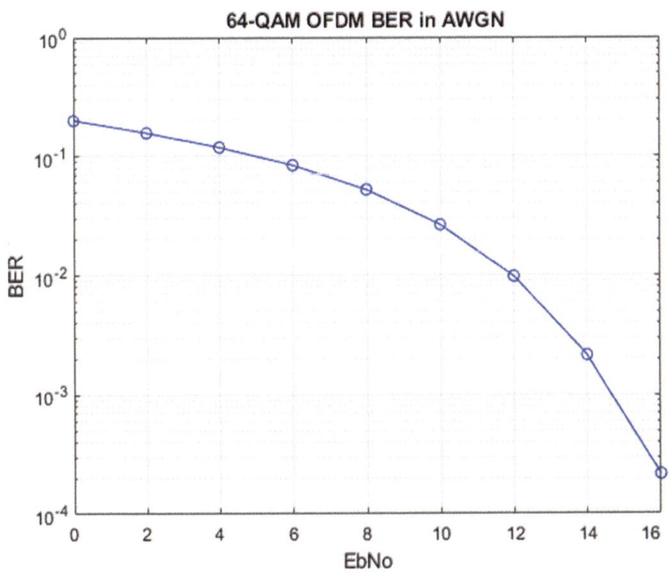

Fig. 4.2 BER of 64-QAM OFDM in AWGN (without CP)

```
%%%%%%%%%%%%%%%%%%%%%%%%%%%%%%%%%%%%%%%%%%%%%%%%%%%%%%%%%%%%%%%
%       FUNCTION TO SIMULATE 64-QAM OFDM IN AWGN CHANNEL (WITHOUT CP)
%                       M is the constellation size
%                   n_bits is the length of binary sequence
%               n_fft is the length of FFT (Fast Fourier Transform)
%         EbNodB is energy per bit to noise power spectral density ratio in dB
%                       ber is output bit error rate
%                       Copyright 2020 RAYmaps
%%%%%%%%%%%%%%%%%%%%%%%%%%%%%%%%%%%%%%%%%%%%%%%%%%%%%%%%%%%%%%%
function[ber]=M_QAM_OFDM(M,n_bits,n_fft,EbNodB)
Es=1;
k=log2(M);
Eb=Es/k;
EbNo=10^(EbNodB/10);
x=transpose(round(rand(1,n_bits)));
y=qammod(x,M,'InputType','bit','UnitAveragePower',true);
n_sym=length(y)/n_fft;

for n=1:n_sym;
  s_ofdm=sqrt(n_fft)*ifft(y((n-1)*n_fft+1:n*n_fft),n_fft);
  wn=randn(1,n_fft)+j*randn(1,n_fft);
  r_ofdm=s_ofdm+sqrt(Eb/(2*EbNo))*wn.';
  s_est((n-1)*n_fft+1:n*n_fft)=fft(r_ofdm,n_fft)/sqrt(n_fft);
end

z=qamdemod(s_est, M,'OutputType','bit','UnitAveragePower',true);
z=z(:);
ber=(n_bits-sum(x==z))/n_bits;
return
%%%%%%%%%%%%%%%%%%%%%%%%%%%%%%%%%%%%%%%%%%%%%%%%%%%%%%%%%%%%%%%
```

4.3 BER of 64-QAM OFDM in Frequency Selective Fading

The real benefits of OFDM become apparent in a frequency selective channel. The introduction of the cyclic prefix (guard period) allows us to remove the intersymbol interference (ISI) in the time domain and frequency domain equalization allows us to overcome the channel variations in the frequency domain.

We consider a simple FIR filter for our channel model with coefficients ht = [0.80 0.54 0.24 0.10 0.04]. This is a simplistic approach since the channel coefficients are all real which means that all multipath components are cophase. To model a more realistic channel we then introduce a uniformly distributed phase shift to all the channel coefficients.

We use an FFT size of 128 and cyclic prefix of 32 samples (16.67 usec) in the simulation given below (Table 4.2).

Table 4.2 LTE physical layer parameters

Transmission BW		1.25 MHz	2.5 MHz	5 MHz	10 MHz	15 MHz	20 MHz
Sub-frame duration		0.5 ms					
Sub-carrier spacing		15 KHz					
Sampling frequency		192 MHz < (1/2 × 3.84 MHz)	3.84 MHz	7.68 MHz (2 × 3.84 MHz)	15.36 MHz (4 × 3.84 MHz)	23 04 MHz (6 × 3.84 MHz)	30.72 MHz (8 × 3.84 MHz)
FFT size		128	256	512	1024	1536	2048
OFDM sym per slot (short/long CP)		7/6					
CP length (usee/ samples)	Short	(4.69.9) × 6, (5.21/ 10) × 1	(4.69/18) × 6, (5.21/20) × 1	(4.69. 36) × 6, (5.21/40) × 1	(4.69/72) × 6, (5.21/80) × 1	(4.69/108) × 6, (5.21/120) × 1	(4.69/144) × 6, (5.21/160) × 1
	Long	(16.67/32)	(16.67/64)	(1667/128)	(1667/256)	(16.67/384)	(16 67/512)

```
%%%%%%%%%%%%%%%%%%%%%%%%%%%%%%%%%%%%%%%%%%%%%%%%%%%%%%%%%%%
%        FUNCTION TO SIMULATE 64-QAM OFDM IN STATIC FREQUENCY SELECTIVE CHANNEL
%                              M is the constellation size
%                       n_bits is the length of binary sequence
%                  n_fft is the length of FFT (Fast Fourier Transform)
%           EbNodB is energy per bit to noise power spectral density ratio in dB
%                              ber is output bit error rate
%                              Copyright 2020 RAYmaps
%%%%%%%%%%%%%%%%%%%%%%%%%%%%%%%%%%%%%%%%%%%%%%%%%%%%%%%%%%%
function[ber]=M_QAM_OFDM_fading(M,n_bits,n_fft,EbNodB)

Es=1;
k=log2(M);
Eb=Es/k;
n_cyc=32;
EbNo=10^(EbNodB/10);

x=transpose(round(rand(1,n_bits)));
y=qammod(x,M,'InputType','bit','UnitAveragePower',true);
n_sym=length(y)/n_fft;

for n=1:n_sym;
  s_ofdm=sqrt(n_fft)*ifft(y((n-1)*n_fft+1:n*n_fft),n_fft);
  s_ofdm_cyc=[s_ofdm(n_fft-n_cyc+1:n_fft); s_ofdm];
  ht=[0.8 0.54 0.24 0.10 0.04];
  ht=ht.*exp(i*2*pi*rand(1,5));
  Hf=fft(ht,n_fft);
  r_ofdm_cyc=conv(s_ofdm_cyc,ht);
  r_ofdm_cyc=(r_ofdm_cyc(1:n_fft+n_cyc));
  wn=sqrt((n_fft+n_cyc)/n_fft)*(randn(1,n_fft+n_cyc)+j*randn(1,n_fft+n_cyc));
  r_ofdm_cyc=r_ofdm_cyc+sqrt(Eb/(2*EbNo))*wn.';
  r_ofdm=r_ofdm_cyc(n_cyc+1:n_fft+n_cyc);
  s_est((n-1)*n_fft+1:n*n_fft)=(fft(r_ofdm,n_fft)/sqrt(n_fft))./Hf.';
end

z=qamdemod(s_est, M,'OutputType','bit','UnitAveragePower',true);
z=z(:);
ber=(n_bits-sum(x==z))/n_bits;
return
%%%%%%%%%%%%%%%%%%%%%%%%%%%%%%%%%%%%%%%%%%%%%%%%%%%%%%%%%%%
```

We also have accounted for the extra energy transmitted for the cyclic prefix in our signal to noise calibration making the BER slightly worse than without this adjustment (Fig. 4.3).

It can be seen that up to 12 dB the BER performance for the two cases is quite similar; however, after 12 dB the BER for the real case drops significantly, whereas the BER for the complex case goes down in linear fashion. The error rate can be significantly improved by employing channel coding and antenna diversity schemes.

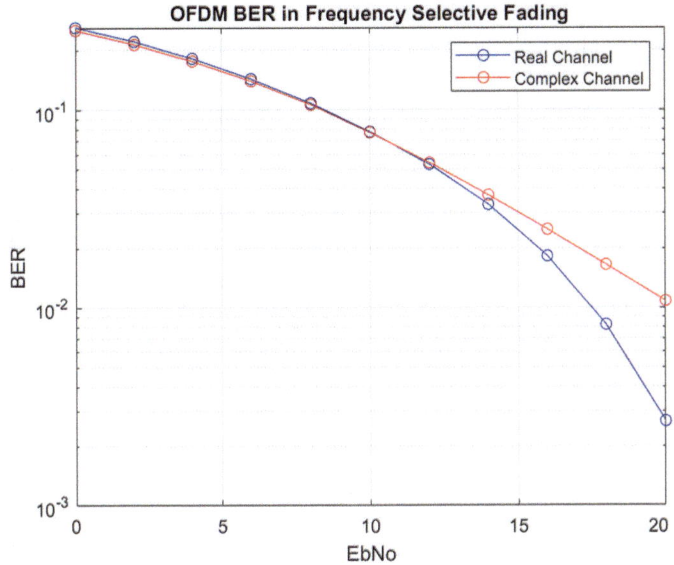

Fig. 4.3 BER of 64-QAM OFDM in fading environment

4.4 BER of 64-QAM OFDM in Frequency Selective Fading-II

In the previous section we had considered a static frequency selective channel. We now consider a time-varying frequency selective channel with 7 taps. Each tap of the time domain filter has a Gaussian distributed real component with variance 1/(2*n_tap) and a Gaussian distributed imaginary component with variance 1/(2*n_tap). The amplitude of each tap is thus Rayleigh distributed, and the phase is uniformly distributed. Since the power in each component is normalized by the filter length (n_tap), the BER performance would remain the same even if the filter length is changed.

```
%%%%%%%%%%%%%%%%%%%%%%%%%%%%%%%%%%%%%%%%%%%%%%%%%%%%%%%
%   FUNCTION TO SIMULATE 64-QAM OFDM IN TIME VARYING FREQUENCY SELECTIVE CHANNEL
%                              M is the constellation size
%                              n_bits is the length of binary sequence
%                              n_fft is the length of FFT (Fast Fourier Transform)
%                EbNodB is energy per bit to noise power spectral density ratio in dB
%                              ber is output bit error rate
%                              Copyright 2020 RAYmaps
%%%%%%%%%%%%%%%%%%%%%%%%%%%%%%%%%%%%%%%%%%%%%%%%%%%%%%%
function[ber]=M_QAM_OFDM_fading(M,n_bits,n_fft,EbNodB)

Es=1;
k=log2(M);
Eb=Es/k;
n_cyc=32;
EbNo=10^(EbNodB/10);
x=transpose(round(rand(1,n_bits)));
y=qammod(x,M,'InputType','bit','UnitAveragePower',true);
n_sym=length(y)/n_fft;
n_tap=7;

for n=1:n_sym;
  s_ofdm=sqrt(n_fft)*ifft(y((n-1)*n_fft+1:n*n_fft),n_fft);
  s_ofdm_cyc=[s_ofdm(n_fft-n_cyc+1:n_fft); s_ofdm];
  ht=(1/sqrt(2))*(1/sqrt(n_tap))*(randn(1,n_tap)+j*randn(1,n_tap));
  Hf=fft(ht,n_fft);
  r_ofdm_cyc=conv(s_ofdm_cyc,ht);
  r_ofdm_cyc=(r_ofdm_cyc(1:n_fft+n_cyc));
  wn=sqrt((n_fft+n_cyc)/n_fft)*(randn(1,n_fft+n_cyc)+j*randn(1,n_fft+n_cyc));
  r_ofdm_cyc=r_ofdm_cyc+sqrt(Eb/(2*EbNo))*wn.';
  r_ofdm=r_ofdm_cyc(n_cyc+1:n_fft+n_cyc);
  s_est((n-1)*n_fft+1:n*n_fft)=(fft(r_ofdm,n_fft)/sqrt(n_fft))./Hf.';
end

z=qamdemod(s_est,M,'OutputType','bit','UnitAveragePower',true);z=z(:);
ber=(n_bits-sum(x==z))/n_bits;
return
%%%%%%%%%%%%%%%%%%%%%%%%%%%%%%%%%%%%%%%%%%%%%%%%%%%%%%%
```

As before, we have used an FFT length of 128 and cyclic prefix of 32 samples. The FFT and IIFT operations are normalized to maintain the signal to noise ratio (SNR). The extra energy transmitted in the cyclic prefix is also accounted for in the SNR calibration.

It is observed that the BER performance of 64-QAM OFDM in the time-varying frequency selective channel is quite similar to that in the static frequency selective channel with complex filter taps (Fig. 4.4). It must be noted that with 64-QAM the goal is to achieve higher bit rate, error rates can be improved using antenna diversity and channel coding schemes.

Given below is the wrapper that should be used along with the above code. The wrapper basically calls the above routine for each value of EbNodB. The constellation size, length of the binary sequence and the FFT size are other inputs to the function. The bit error rate at the specific EbNodB is the output of the function.

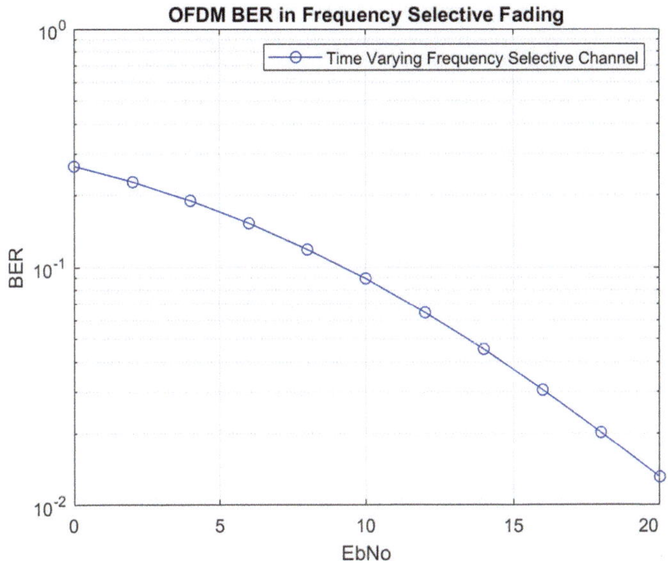

Fig. 4.4 BER of 64-QAM OFDM in time-varying frequency selective channe

```
%%%%%%%%%%%%%%%%%%%%%%%%%%%%%%%%%%%%%%%%%%%%%%%
clear all
close all

M=64;
k=log2(M);
n_fft=128;
l=k*n_fft*1e4;
EbNodB=0:2:20;
for n=1:length(EbNodB);n
  ber(n)=M_QAM_OFDM_fading(M,l,n_fft,EbNodB(n));
end;

semilogy(EbNodB,ber,'bo-');
grid on
xlabel('EbNo')
ylabel('BER')
title('OFDM BER in Frequency Selective Fading')
%%%%%%%%%%%%%%%%%%%%%%%%%%%%%%%%%%%%%%%%%%%%%%%
```

In the future we would use the standard LTE channel models, namely EPA, EVA and ETU in our simulation.

4.5 Can We Do Without a Cyclic Prefix

Have you ever thought that cyclic prefix in OFDM is just a gimmick, and we could do equally well by using a guard period, i.e., a period of no transmission between two OFDM symbols? Well, one way to find out if this is true is by running a bit error rate simulation with and without a cyclic prefix (only a vacant guard period). We use the 64-QAM OFDM simulation that we developed previously. The channel is modeled as 7-tap FIR filter with each tap having a Rayleigh distribution.

We simulate the case of a vacant guard period by inserting zeros in the time slot dedicated for the CP (32 samples). It is observed that there is a vast difference in the bit error rate (BER) for the two cases (Fig. 4.5). In fact, in the case of no CP the BER hits an error floor at around 20 dB. Increasing the signal to noise ratio does not improve the BER performance any further.

Now to answer the question "why does the CP work" we have to revisit the concept of circular convolution from our DSP course. It is well known that performing circular convolution of two sequences in the time domain is equivalent to multiplication of their DFT's in the frequency domain. So, if a wireless channel performed circular convolution, we could do simple division to recover the signal after the FFT operation in the receiver.

$$Y(k) = X(k)H(k)$$
$$X(k) = Y(k)/H(k)$$

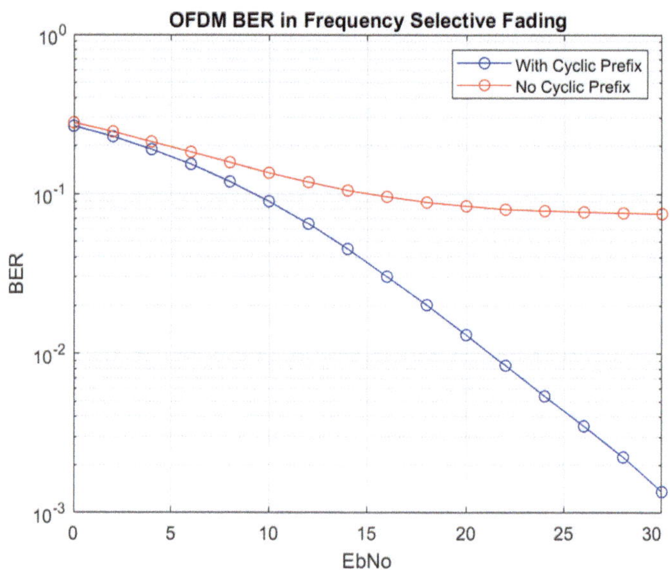

Fig. 4.5 BER with and without cyclic prefix

But the wireless channel does not perform circular convolution, it performs linear convolution. So, the trick is to make this linear convolution appear as circular convolution by appending a cyclic prefix. The result is that equalization can be performed at the receiver by simple division.

For a more elaborate discussion on circular convolution you may visit CP-1 and for details about how CP is used in OFDM you may visit CP-2.

Note: In an actual system there would be AWGN noise added to the received signal as well, giving us the following relationships.

$$Y(k) = X(k)H(k) + \mathrm{W(k)}$$

$$\hat{X}(k) = \frac{Y(k)}{H(k)} = X(k) + \frac{W(k)}{H(k)}$$

4.6 Peak to Average Power Ratio (PAPR)

Peak to Average Power Ratio (PAPR) as the name suggests is the ratio of peak signal power to the average signal power and has received considerable attention in the context of multicarrier signals like OFDM which exhibit a high PAPR. The downside of this high PAPR is that the power amplifier in the transmitter is operated at a relatively lower power level so that the peaks in the signal are not distorted by the saturating amplifier. This is called the amplifier backoff, and it plays an important part in wireless system design [1].

The reason for this high PAPR is that when multiple sinusoids are added together in a multicarrier transmission the resulting signal exhibits constructive and destructive behavior. The higher the number of these sinusoids higher is the PAPR. The figure below illustrates this behavior.

It is observed that the PAPR of a signal composed of two sinusoids is greater than that of a single sinusoid. Similarly, the PAPR of a signal composed of three carriers is even higher. The PAPR for the case of a single tone, two tones and three tones is 2.00, 3.10 and 4.15, respectively. Or on a logarithmic scale it can be expressed as 3.01 dB, 4.91 dB and 6.18 dB, respectively. So, if the power amplifier in a wireless system starts saturating at 24 dBm, then the average signal power of three tone system must not exceed 24–6.18 = 17.82 dBm. In practical systems techniques are adopted that decrease the PAPR so that the power amplifier can operate close to its maximum limit. One simple technique clips the peaks in the signal, while another adds tones at the unused frequencies such that the total PAPR is reduced (Figs. 4.6 and 4.7).

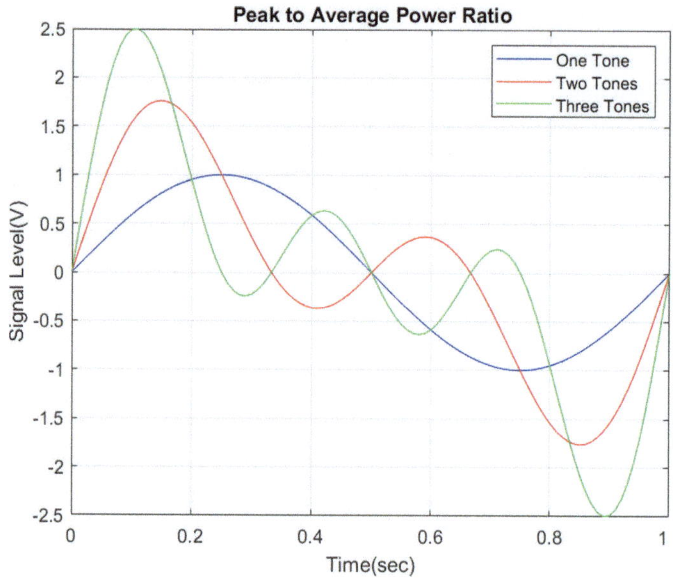

Fig. 4.6 Constructive and destructive behavior of multiple carriers

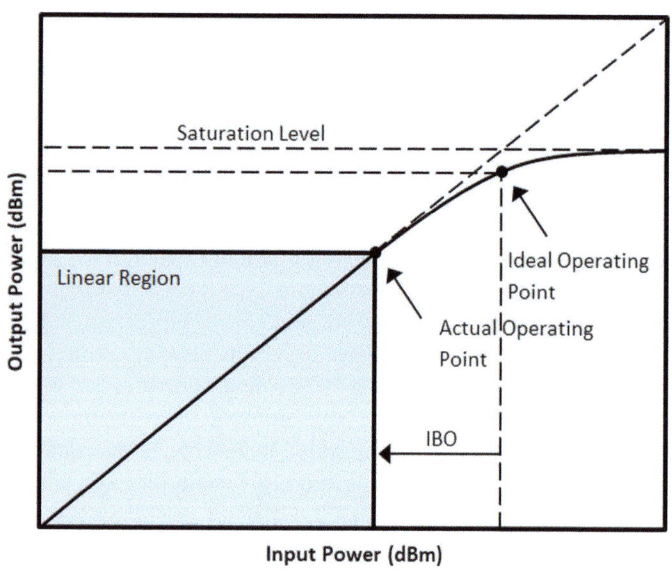

Fig. 4.7 Power amplifier input and output behavior

Note:

1. The three tones in the above example are at 1, 2 and 3 Hz.
2. Ideal operating point is where the amplifier achieves maximum efficiency. Amplifiers often operate most efficiently just into saturation [1].
3. 1 dB compression point, or P1dB, is the power level where the output power of a device, such as an amplifier, starts to drop off from the linear input/output power curve and reaches a point where the actual output power is 1 dB less than the theoretical linear curve [2].
4. IBO or Input Back Off is the reduction in input so that amplifier stays within the linear region.
5. Sometimes it's more important to look at distribution of the signal rather than a peak value; e.g., if the signal peak goes into non-linear region 0.001% of the time then maybe we do not need to worry that much.
6. Another important metric that is used to measure the peakedness of signal is the cubic metric (CM).

Questions and Numerical Problems

1. What are some of the common multiplexing and multiple access techniques used in wireless communications?
2. What is the difference between OFDM and OFDMA?
3. What is the meaning of Orthogonal in Orthogonal Frequency Division Multiplexing?
4. What is the typical subcarrier spacing used in 4G and 5G? What will happen if the subcarrier spacing is further reduced?
5. Why is cyclic prefix used in OFDM? What are the advantages and disadvantages (if any)?
6. Is there any difference in bit error rate performance of 64-QAM OFDM and 256-QAM OFDM in (a) AWGN (b) Rayleigh fading?
7. Is there any advantage of using OFDM in a flat fading scenario? Explain.
8. What is the difference between a channel with seven taps and nine taps? What type of channel is simulated using just one tap?
9. Why is the bit error rate performance better when we use purely real channel coefficients in our simulation?
10. We know that in OFDM, IFFT operation is performed at the transmitter and FFT operation is performed at the receiver, can this be reversed, i.e., FFT at the transmitter and IFFT at the receiver?
11. Why can't we use a guard period instead of a cyclic prefix in OFDM? Explain mathematically.
12. In which scenarios is a short cyclic prefix used and in which scenarios is a long cyclic prefix more suitable?
13. Calculate the Peak to Average Power Ratio for a signal composed of (a) Ten tones (b) Twenty tones (c) Fifty tones, using simulation. Comment on how this impacts a practical multicarrier system.
14. What is Cube Metric, and how is it used to measure the peakedness of a signal?

15. We mostly consider only linear effects in our simulation of wireless communication systems. What are some of the common non-linearities that a wireless system design engineer may face?

Useful Links

1. BER of 64-QAM OFDM in AWGN
   https://www.raymaps.com/index.php/ber-of-64-qam-ofdm-in-awgn/
2. BER of 64-QAM OFDM in Frequency Selective Fading
   https://www.raymaps.com/index.php/ber-of-64-qam-ofdm-in-fading/
3. BER of 64-QAM OFDM in Frequency Selective Fading-II
   https://www.raymaps.com/index.php/ber-of-64-qam-ofdm-in-frequency-selective-fading-ii/
4. Can We Do Without a Cyclic Prefix?
   https://www.raymaps.com/index.php/can-we-do-without-a-cyclic-prefix/
5. Peak to Average Power Ratio (PAPR)
   https://www.raymaps.com/index.php/peak-to-average-power-ratio-papr/

References

1. https://www.microwavejournal.com/articles/31597-using-rf-power-meters-for-papr-analysis-and-reduction
2. https://ez.analog.com/rf/f/q-a/71408/what-is-analog-devices-definition-of-output-p1db

Chapter 5
Shannon Capacity

5.1 Introduction

In this chapter we discuss the capacity of wireless communication systems which is fundamental to how they are designed. We learnt about modulation and coding techniques in Chap. 2 but what can be practically achieved with these techniques is discussed in this chapter.

It was in 1948 that Claude Shannon gave his theory known as "A Mathematical Theory of Communication" which formed the basis of the field of Information Theory. He postulated that whatever the source of information, a human voice, a piece of music or a text document, must first be converted into bits for efficient transmission over the medium. He along with John Tukey is credited with coining the term "bit" which is now the most commonly used term in communication theory.

So what are the main contributions of Claude Shannon's seminal work that he presented about 70 years back? There are three main contributions:

i. First of all he introduced the term Entropy Rate (H) which is the minimum number of bits per second required to represent the information to be transferred. Higher the Entropy Rate the more difficult it is to transfer the information.
ii. He then proposed the capacity theorem which can be used to calculate the maximum number of bits per second that can be reliably communicated in the presence of noise in the channel. This is known as capacity (C) of the channel which depends upon bandwidth of the channel and the signal to noise ratio (SNR).
iii. Finally he showed that reliable communications is only possible if H < C; i.e., Entropy Rate is less than capacity. To explain this concept, think of H as rate of flow of water and C as capacity of the pipe, which depends upon its cross section. We know that the rate of flow of water cannot be greater than the capacity of the pipe.

Y. Ahmed, *Recipes for Communication and Signal Processing*,
Signals and Communication Technology,
https://doi.org/10.1007/978-981-99-2917-7_5

Initially the capacity theorem was presented for an Additive White Gaussian Noise (AWGN) channel but later it was adapted for fading channels and more recently to multiinput and multioutput (MIMO) channels. In this chapter we focus on AWGN case and give examples of GSM, LTE and 5G standards and calculate their capacities. We also do a comparison of OFDMA and CDMA capacities and show that dividing a wide channel into narrow frequency bins does not change the capacity. Finally we briefly touch upon capacity calculation for SISO and MIMO fading channels.

Two of the main ingredients which allow us to achieve Shannon Capacity in 5G systems are LDPC codes (already discussed in Chap. 2) and polar codes. LDPC codes are used for user data while polar codes are used for control information. Polar codes are especially useful for short block lengths and have relatively simple encoding and decoding.

5.2 Shannon Capacity of GSM in an AWGN Channel

We know that GSM bit rates can vary from a few kbps to a theoretical maximum of 171.2 kbps (GPRS). But what is the actual capacity of a 200 kHz GSM channel. We can use the Shannon Capacity theorem to find this capacity. We know from Shannon Capacity theorem that:

$$C = B\log_2(1 + SNR)$$

or

$$C = B\log_2\left(1 + \frac{P}{N}\right)$$

or

$$C = B\log_2\left(1 + \frac{P}{BN_o}\right).$$

Here P is the signal power in Watts, N is the noise power in Watts, N_o is the noise power spectral density in Watts/Hz and B is the bandwidth in Hz.

The noise power can be found by using the following formula, where k is the Boltzmann constant in Joules/Kelvin, T is the temperature in Kelvin and B is the bandwidth in Hz.

$$\mathrm{N} = \mathrm{kTB} = (1.38\mathrm{e} - 23) * (293) * (200\mathrm{e}3) = 8.08\mathrm{e} - 16\mathrm{W} = -121\mathrm{dBm}.$$

Let us now assume a signal power of – 90dBm. This gives us an SNR of 31 dB or 1258.9 on linear scale. The capacity can thus be calculated as:

$$\mathrm{C} = 200\mathrm{e}3 * \log 2(1 + 1258.9) = 2.06\,\mathrm{Mbps}.$$

This is the capacity if all time slots are allocated to a single user. If only one time slot is allocated to a user, the capacity would be reduced to 257.48 kbps.

Note: One way to look at it is that each user is effectively experiencing a bandwidth of 25 kHz only. So keeping the SNR to be the same the capacity is divided by eight.

5.3 Shannon Capacity of GSM in a Fading Channel

In the previous section we calculated the Shannon Capacity of a 200 kHz GSM channel in an AWGN (Additive White Gaussian Noise) environment. However, in a practical scenario the capacity is limited by time-varying fading and interference. Let us consider a fading channel with four possible states corresponding to SNRs of 15, 10, 5 and 0 dB. The probability of these states is estimated to be 0.50, 0.25, 0.15 and 0.10, respectively. The Shannon Capacity of such a channel is given as (assuming that the channel state information is known at the receiver):

$$C = \sum_{i=1}^{4} B \log_2(1 + \gamma_i) p(\gamma_i),$$

where

C is the capacity in bps
B is the bandwidth in Hertz
γ_i is the signal to noise ratio of the *ith* state and
$p(\gamma_i)$ is the probability of the *ith* state.

The above equation can be rewritten as:

$$C = B \sum_{i=1}^{4} \log_2(1 + \gamma_i) p(\gamma_i)$$

$$C = 200000\big(\log_2(1 + 31.62)0.50 + \log_2(1 + 10.00)0.25 + \log_2(1 + 3.16)0.15 + \log_2(1 + 1)0.10\big)$$

$$C = 757.44\,\mathrm{kbps}.$$

Assuming that only one out of eight time slots is allocated to any user the Shannon Capacity of a GSM channel is reduced to $757.44/8 = 94.68$ kbps.

Note:

1. The contribution of the high SNR states dominates the capacity of the channel. The contribution of the four states in terms of percentage capacity is given as 66.38, 22.84, 8.14 and 2.64%. So approximately 90% of the capacity can be achieved if only two of the states are used for transmission and remaining two states are completely switched off.
2. In a practical scenario the signal to noise ratio is time varying and continuous, it cannot be quantized into four states. In such a case integration would have to be performed instead of summation, over the range of SNRs.
3. As per calculations above, spectral efficiency of more than 3.5 bits/sec/Hz can be achieved but this is far away from actual spectral efficiency of GSM (0.5–1.0).

5.4 Shannon Capacity of LTE

Shannon Capacity of LTE in AWGN can be calculated by using the Shannon Capacity formula:

$$C = B\log_2(1 + SNR)$$

or

$$C = B\log_2\left(1 + \frac{P}{BN_o}\right)$$

As before, the signal power P is set at −90 dBm, the noise power spectral density N_o is set at 4.04e-21 W/Hz (−174 dBm/Hz) and the bandwidth is varied from 1.25 to 20 MHz (Fig. 5.1).

It is seen that the capacity increases from about 10 Mbps to above 70 Mbps as the bandwidth is varied from 1.25 to 20 MHz (keeping the signal power constant). It is seen that capacity does not increase linearly with bandwidth. For a bandwidth increase of 16X the capacity only increases sevenfold. This is because with increase in bandwidth the noise power also increases and SNR decreases (remember that noise is white that is spread uniformly in the frequency domain).

It must be noted that this is the capacity with a single transmit and single receive antenna (MIMO capacity would obviously be higher).

5.5 5G Data Rates and Shannon Capacity

Recently I came across a post from T-Mobile in which they claim to have achieved a download speed of 5.6 Gbps over a 100 MHz channel resulting in a spectral efficiency of more than 50 bps/Hz. This was achieved in an MU-MIMO configuration with

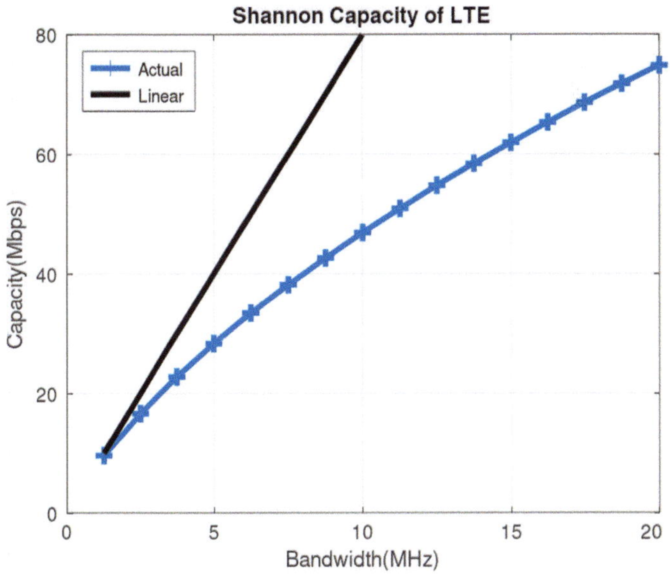

Fig. 5.1 Shannon capacity of LTE as a function of bandwidth

eight connected devices having an aggregate of 16 parallel streams, i.e., two parallel streams per device. The channel used for this experiment was the mid-band frequency of 2.5 GHz [1].

Now let us revisit the Shannon Capacity theorem and see what data rate it predicts with the above parameters. Shannon Capacity theorem for a MIMO System with N parallel streams is given as [2]:

$$C = NB\log_2(1 + SNR),$$

where

$N = 16$	is the number of parallel streams enabled by multiple antennas at Tx and Rx
$B = 100$ MHz	is the total bandwidth available to the carrier in the 2.5 GHz band
$SNR = 10$	is the signal to noise ratio (a moderately good SNR is assumed).

Plugging these numbers in, we get a capacity limit of 5.54 Gbps, approximately the same number obtained in the experiment. But it all depends upon the signal to noise ratio that has not been mentioned by T-Mobile in the reference article. With an SNR = 20 dB, a very good channel condition, the capacity increases to about 10.65 Gbps but with an SNR = 0 dB, a likely scenario at the cell edge, the capacity drops to 1.6 Gbps. The details for the capacity at different SNRs are given in the table and figure (Fig. 5.2; Table 5.1).

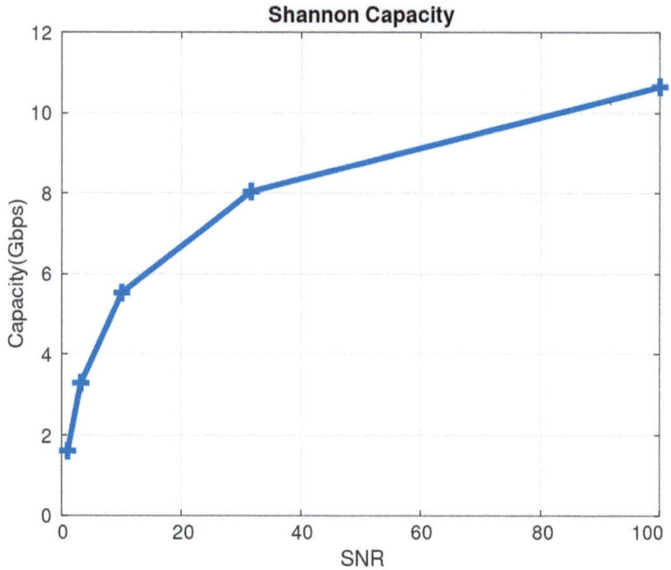

Fig. 5.2 5G Shannon capacity as a function of SNR

Table 5.1 5G Shannon capacity as a function of SNR

SNR (dB)	SNR	Streams	BW (MHz)	Capacity (Gbps)
0	1.00	16	100	1.60
5	3.16	16	100	3.29
10	10.00	16	100	5.54
15	31.62	16	100	8.04
20	100.00	16	100	10.65

Please note that the above capacity calculations are for eight users simultaneously connected over the 100 MHz channel. The data rate achieved by a single user, in the T-Mobile experiment, was 5600/8 = 700 Mbps.

Also note that much higher bandwidths are available in the millimeter wave band, and this may increase the data rates tenfold or even higher.

5.6 Narrowband Versus Wideband

Somebody recently asked me this question "*Does Shannon Capacity Increase by Dividing a Frequency Band into Narrow Bins.*" To be honest I was momentarily confused and thought that this may be the case since many of the modern Digital Communication Systems do use narrow frequency bins, e.g., LTE and 5G. But on closer inspection I found that the Shannon Capacity does not change, in fact it remains exactly the same. Following is the reasoning for that.

Shannon Capacity is calculated as:

$$C = B\log_2(1 + SNR)$$

or

$$C = B\log_2\left(1 + \frac{P}{BN_o}\right).$$

Now if the bandwidth B is divided into 10 equal blocks then the transmit power P for each block would also be divided by 10 to keep the total transmit power for the entire band to be constant. This means that the factor P/BN_o remains constant. So, the total capacity for the 10 blocks would be calculated as:

$$C = 10(B/10)\log_2\left(1 + \frac{(P/10)}{(B/10)N_o}\right).$$

So, the Shannon Capacity for the entire band remains the same.

PS: The reason for the narrower channels is that for a narrow channel the channel appears relatively flat in the frequency domain and the process of equalization is thus simplified (a simple multiplication or division would do).

Note: N_o is the noise power spectral density and BN_o is the noise power.

5.7 Shannon Capacity CDMA Versus OFDMA

We have previously discussed Shannon Capacity of CDMA and OFMDA; here we will discuss it again in a bit more detail. Let us assume that we have 20 MHz bandwidth for both the systems which is divided among u users. For OFDMA we assume that each user gets 20/u MHz bandwidth and there are no guard bands or pilot carriers. For CDMA we assume that each user utilizes full 20 MHz bandwidth and users are separated by their unique code. We can say that for OFDMA each user has a dedicated channel, whereas for CDMA the channel is shared between u simultaneous users.

We know that Shannon Capacity is given as:

$$C = B\log_2(1 + SNR)$$

or in the case of CDMA

$$C = B\log_2(1 + SINR),$$

where B is the bandwidth and SINR is the signal to noise plus interference ratio. For OFDMA the SNR is given as

$$SNR = \frac{P_u}{BN_o},$$

where P_u is the signal power of a single user and N_o is the noise power spectral density. For CDMA the calculation of SINR is a bit more complicated as we have to take into account the Multiple Access Interference. If the total number of users is u the SINR is calculated as

$$SINR = \frac{P_u}{BN_o + (u-1)P_u}$$

Please note that each user transmits a power P_u and therefore total power transmitted is uP_u. The code below plots the capacities of CDMA and OFDMA systems as a function of the number of users.

```
%%%%%%%%%%%%%%%%%%%%%%%%%%%%%%%%%%%%%%%%%%%%%%%%%%%%%%%
%                CAPACITY OF CDMA and OFDMA
%                u is the number of users
%                Pu is power of a single user
%                No is the Noise Power Spectral Density
%                B is the bandwidth
%
%                Copyright 2020 RAYmaps
%%%%%%%%%%%%%%%%%%%%%%%%%%%%%%%%%%%%%%%%%%%%%%%%%%%%%%%
clear all
close all

u=1:25;
Pu=1e-14;
No=(1.38e-23)*(293);
B=20e6;

C_CDMA=u.*(B*log2(1+Pu./(B*No+(u-1)*Pu)));

B=B./u;
C_OFDMA=u.*(B.*log2(1+(Pu./(B*No))));

plot(u,C_CDMA/1e6, 'bo-');
hold on
plot(u,C_OFDMA/1e6,'ro-');
hold off
title('CDMA vs OFDMA Capacity')
xlabel('Number of Users')
ylabel('Capacity(Mbps)')
legend('CDMA','OFDMA')
grid on
%%%%%%%%%%%%%%%%%%%%%%%%%%%%%%%%%%%%%%%%%%%%%%%%%%%%%%%
```

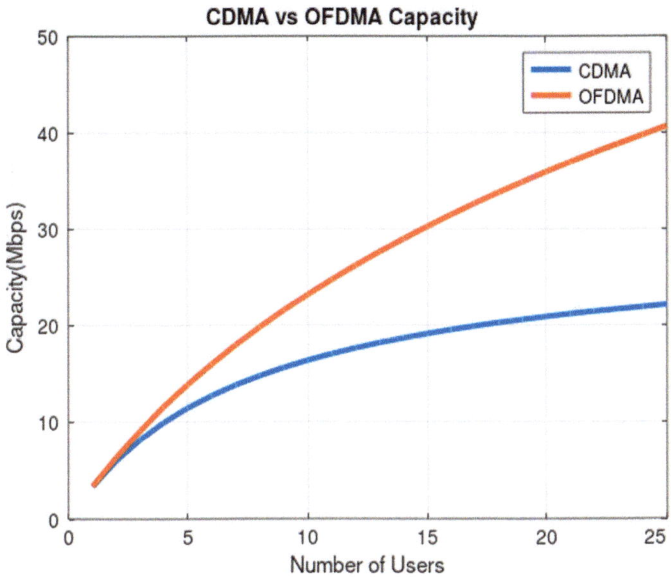

Fig. 5.3 Shannon capacity of CDMA and OFDMA

We see that the capacity of OFDMA increases much more rapidly as number of users increases. This is mainly due to increase in signal to noise ratio as each new user gets full transmit power of P_u watts and noise power decreases as user bandwidth decreases (remember noise power is given as BN_o). If we increase the power of each user in simulation, we see that it further improves the capacity of OFDMA but capacity of CDMA system does not improve much as Multiple Access Interference increases. In fact CDMA capacity hits a ceiling at about 30 Mbps or spectral efficiency of 1.5 bits/sec/Hz as per user power is increased (Fig. 5.3).

5.8 MIMO Capacity in a Fading Environment

The Shannon Capacity of a channel is the data rate that can be achieved over a given bandwidth (BW) and at a particular signal to noise ratio (SNR) with diminishing bit error rate (BER). This has been discussed in earlier sections for the case of SISO channel and Additive White Gaussian Noise (AWGN). For a MIMO fading channel the capacity with channel not known to the transmitter is given as (both sides have been normalized by the bandwidth [3]):

$$C = \log_2\left[\det\left(I_{N_R} + \frac{\gamma}{N_T} H H^H\right)\right],$$

where N_T is the number of transmit antennas, N_R is the number of receive antennas, γ is the signal to interference plus noise ratio (SINR), I_{NR} is the $N_R \times N_R$ identity matrix and H is the $N_R \times N_T$ channel matrix. Furthermore, h_{ij}, an element of the matrix H defines the complex channel coefficient between the ith receive antenna and jth transmit antenna. It is quite obvious that the channel capacity (in bits/sec/Hz) is highly dependent on the structure of matrix H. Let us explore the effect of H on the channel capacity.

Let us first consider a 4×4 case ($N_T = 4$, $N_R = 4$) where the channel is a simple AWGN channel and there is no fading. For this case $h_{ij} = 1$ for all values of i and j. It is found that channel capacity of this simple channel for an SINR of 10 dB is 5.36 bits/sec/Hz. It is further observed that the channel capacity does not change with number of transmit antennas and increases logarithmically with increase in number of receive antennas. Thus, it can be concluded that in an AWGN channel no multiplexing gain is obtained by increasing the number of transmit antennas.

We next consider a more realistic scenario where the channel coefficients h_{ij} are complex with real and imaginary parts having a Gaussian distribution with zero mean and variance 0.5 per dimension. Since the channel H is random, the capacity is also a random variable with a certain distribution. An important metric to quantify the capacity of such a channel is the complimentary cumulative distribution function (CCDF). This curve basically gives the probability that the MIMO capacity is above a certain threshold.

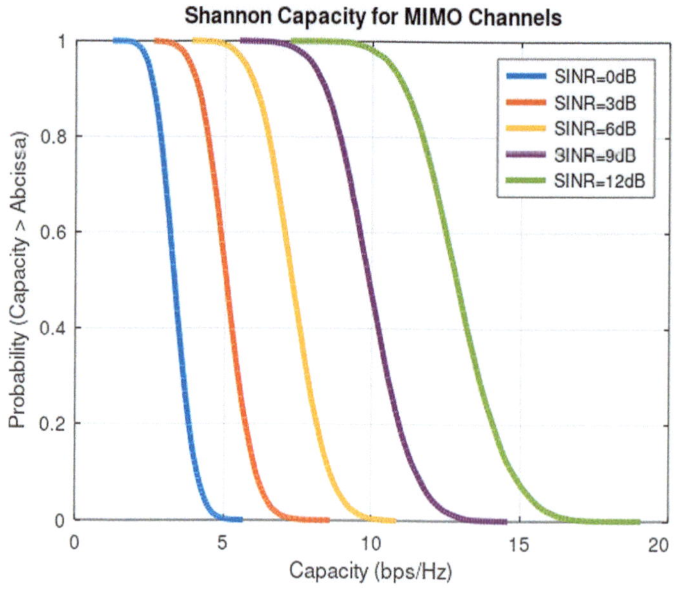

Fig. 5.4 Complimentary cumulative distribution function of capacity

It is obvious (Fig. 5.4) that there is a very high probability that the capacity obtained for the MIMO channel is significantly higher than that obtained for an AWGN channel, e.g., for an SINR of 9 dB there is 90% probability that the capacity is greater than 8 bps/Hz. Similarly, for an SINR of 12 dB there is a 90% probability that the capacity is greater than 11 bps/Hz. For a stricter threshold of 99% the above capacities are reduced to 7.2 and 9.6 bps/Hz.

In a practical system the channel coefficients h_{ij} would have some correlation which would depend upon the antenna spacing. Lower the antenna spacing higher would be the antenna correlation and lower would be the MIMO system capacity. This would be discussed in a future post.

The MATLAB code for calculating the CCDF of channel capacity of a MIMO channel is given below.

```
%%%%%%%%%%%%%%%%%%%%%%%%%%%%%%%%%%%%%%%%%%%%%%%%%%%%%%%%%%%%%%%%%%
%          MIMO CAPACITY IN A FADING ENVIRONMENT
%              Nr is the number of receive antennas
%              Nt is the number of transmit antennas
%                    H is the channel matrix
%   C is the capacity for the realization of a particular channel H
%                   Copyright 2020 RAYmaps
%%%%%%%%%%%%%%%%%%%%%%%%%%%%%%%%%%%%%%%%%%%%%%%%%%%%%%%%%%%%%%%%%%
clear all
close all

Nr=4;
Nt=4;
I=eye(Nr);
SINR=0;
g=10^(SINR/10);

for n=1:10000
  H=sqrt(1/2)*randn(Nr,Nt)+j*sqrt(1/2)*randn(Nr,Nt);
  C(n)=log2(det(I+(g/Nt)*(H*H')));
end

[a,b]=hist(real(C),100);
a=a/sum(a);
plot(b,1-cumsum(a));
xlabel('Capacity (bps/Hz)')
ylabel('Probability (Capacity>Abcissa)')
grid on
%%%%%%%%%%%%%%%%%%%%%%%%%%%%%%%%%%%%%%%%%%%%%%%%%%%%%%%%%%%%%%%%%%
```

Ergodic Capacity: Some Important Cases

i. **MIMO Capacity (CSI available at receiver)**

$$C = \log_2 \det\left(I_{N_R} + \frac{\gamma}{N_T} H H^H\right)$$

$$C = \sum_{i=1}^{r} \log_2\left(1 + \frac{\gamma}{N_T}\lambda_i\right)$$

λ_i is the i th singular value or eigenvalue of HH^H
r is the rank of H, $r = \min(N_T, N_R)$

ii. **SIMO Capacity (CSI available at receiver)**

$$C = \sum_{i=1}^{r} \log_2\left(1 + \tfrac{\gamma}{N_T}\lambda_i\right)$$
$$C = \log_2\left(1 + \tfrac{\gamma}{N_T}\lambda_1\right)$$
$$C = \log_2(1 + \gamma N_R)$$
$$\lambda_1 = \|h\|^2 = N_R,\ N_T = 1$$

iii. **MISO Capacity (CSI available at receiver)**

$$C = \sum_{i=1}^{r} \log_2\left(1 + \tfrac{\gamma}{N_T}\lambda_i\right)$$
$$C = \log_2\left(1 + \tfrac{\gamma}{N_T}\lambda_1\right)$$
$$C = \log_2(1 + \gamma)$$
$$\lambda_1 = \|h\|^2 = N_T$$

The Frobenius norm $\|.\|$ requires that we cycle through all matrix entries, add their squares and then take the square root.

Another point to be noted here is that capacity discussed here is ergodic capacity, where it is assumed that the channel transitions over all the fading states. In simple words it means that to find the total capacity we take the average of the capacities found for different channel realizations.

Questions and Numerical Problems

1. How is noise power calculated? Give the mathematical formula and explain with the help of an example.
2. Where in the wireless system is AWGN noise added?
3. Calculate the Shannon Capacity of a system with 20 MHz bandwidth and signal to noise ratio of 10 dB.
4. Calculate the Shannon Capacity of the above system when the bandwidth is doubled but signal to noise ratio is halved (on linear scale). What is the difference in capacity compared to Part 3?

5. Calculate the Shannon Capacity of the above system when the bandwidth is halved but signal to noise ratio is doubled (on linear scale). What is the difference in capacity compared to Part 3?
6. A Rayleigh fading channel has four states with equal probability. The signal to noise ratio for the four states is 5, 10, 15 and 20 dB. The bandwidth is fixed at 20 MHz. What is the Shannon Capacity of the system? What is the difference in capacity compared to Part 3?
7. A Rayleigh fading channel has four states with probabilities 0.1, 0.4, 0.3, 0.2. The signal to noise ratio for the four states is 5 dB, 10 dB, 15 dB and 20 dB, respectively. The bandwidth is fixed at 20 MHz. What is the Shannon Capacity of the system? What is the difference in capacity compared to Part 6?
8. Explain what is meant by the soft capacity of CDMA?
9. What is the fundamental reason that OFDMA has replaced CDMA as the multiple access scheme in modern wireless communication systems?
10. Explain the role of channel coding in achieving Shannon Capacity in modern wireless communication systems.
11. In the future wireless communication systems what will be the biggest impact of using millimeter wave and terahertz frequencies?
12. In the future wireless communication systems what will be the impact of using large antenna arrays?
13. Give the formula for Shannon Capacity for a MIMO system operating in a fading environment. Explain its various components.
14. Please explain how MIMO capacity can be higher in a fading environment than in an AWGN environment?
15. Please explain if the channel coefficients h_{ij} as described above, have some correlation, how would it impact channel capacity?

Useful Links

1. Shannon Capacity of GSM in an AWGN Channel
https://www.raymaps.com/index.php/shannon-capacity-of-a-gsm-channel-mbps/
2. Shannon Capacity of GSM in a Fading Channel
https://www.raymaps.com/index.php/shannon-capacity-of-a-gsm-channel-in-fading-environment/
3. Shannon Capacity of LTE
https://www.raymaps.com/index.php/shannon-capacity-of-lte-ideal/
4. 5G Data Rates and Shannon Capacity
https://www.raymaps.com/index.php/5g-data-rates-and-shannon-capacity/
5. Narrowband versus Wideband
https://www.raymaps.com/index.php/does-shannon-capacity-increase-by-dividing-a-frequency-band-into-narrow-bins/

6. Shannon Capacity CDMA versus OFDMA
   https://www.raymaps.com/index.php/shannon-capacity-cdma-vs-ofdma/
7. MIMO Capacity in a Fading Environment
   https://www.raymaps.com/index.php/mimo-capacity-fading-environment/

References

1. https://www.t-mobile.com/news/network/t-mobile-achieves-mind-blowing-5g-speeds-with-mu-mimo
2. https://www.waveform.com/a/b/guides/5g-and-shannons-law
3. Foschini, G.J., Gans, M.J.: On limits of wireless communications in a fading environment when using multiple antennas. Wireless Personal Commun. 6:311–335

Chapter 6
Antenna Arrays

6.1 Introduction

The simplest form of an antenna is an isotropic point source that radiates uniformly in all directions. The power density at d meters from the point source is given as $P/4\pi d^2$, i.e., total transmit power divided by the surface area of the sphere with radius d. To capture this power at a large distance from the source we need a sizable antenna aperture. The antenna aperture is directly proportional to the gain of the antenna and squared of the wavelength. You might recall that in the Friis equation, for calculation of received power in free space, there is a squared wavelength term in there. It basically stems from the relationship with aperture of the receive antenna. It's like catching rain with a bucket, the larger the bucket (antenna aperture) the more water (energy) you will collect.

Real antennas are never isotropic (sun is an example of an isotropic radiator) rather they radiate more in one direction and less in others. The quantity used to describe the directionality of an antenna is called the gain or directivity. Although both the terms are used interchangeably, there is a small difference. Antenna gain accounts for the efficiency of the antenna, while directivity does not. Mathematically, the gain of the antenna is given as the product of directivity with efficiency.

$$G(\theta, \phi) = ED(\theta, \phi)$$

Antenna gain or directivity is normally given with reference to an isotropic source, but sometimes it may be given with reference to a dipole and expressed as dBd. Another important characteristic of the antenna is its impedance. A simple half-wave dipole has an impedance of about 75 ohms, same as a coaxial cable. A good antenna must act as a transducer between the RF front end of the transmitter or receiver and the free space. The input to the transmitting antenna is an electrical signal coming out of an RF amplifier, and output is an electromagnetic wave. The reverse is true for a receiving antenna. The impedance of free space is usually assumed to be 377 ohms

Y. Ahmed, *Recipes for Communication and Signal Processing*,
Signals and Communication Technology,
https://doi.org/10.1007/978-981-99-2917-7_6

($Z = E/H$), but this is only true in the far field of an antenna, typically a distance of more than one wavelength from the antenna. At smaller distances the impedance is much higher, and the relationship between the E-field and H-field is complex.

In modern wireless communications, single antenna elements are rarely used, they are usually combined in the form of arrays. The most common array geometries are uniform linear array, non-uniform linear array, rectangular array, and circular array. The behavior of an array is defined by the array factor which depends upon the interelement spacing, number of elements in the array and arrangement of the array. The antenna radiation pattern can be calculated by multiplying the array factor with the Element Factor of a given antenna. One important design rule that must be remembered is that interelement spacing must be less than half the wavelength to avoid grating lobes. But we know from the theory of MIMO communications that interelement spacing must not be less than half the wavelength to avoid high correlation. So these are somewhat conflicting requirements.

Higher the number of antennas greater is the gain or directivity of the antenna, and it's easier to resolve two closed spaced users in the angular domain. The most commonly used antenna configurations in modern wireless communications are square (a special case of rectangular array) or circular. Just to give an example an 8×8 square array with 0 dB gain of each element has a combined gain of 10 * log10(64) = 18 dB. If the gain of each individual element is 3 dB, the total gain of the array increases to 21 dB. In the future it is envisaged that Massive MIMO Systems would have hundreds of antennas at the transmitter and receiver. Smaller wavelengths at millimeter wave and terahertz frequencies would allow for realistic size of the extremely large antenna arrays.

6.2 Fundamentals of a Uniform Linear Array (ULA)

A uniform linear array (ULA) is a collection of sensor elements equally spaced along a straight line. The most common type of sensor is a dipole antenna that can transmit and receive electromagnetic waves over the air. Other types of sensors include acoustic sensors that may be used in air or under water. The requirements of a ULA are different for different applications, but the most common requirement is to improve the signal to noise ratio (SNR) and to improve its response (gain) in a particular direction. The second property means that the array accepts a signal from a particular direction and rejects the signal from another direction just as required in radar.

The graphical representation and the mathematical framework for the problem are shown below. It is assumed that electromagnetic waves (rays) arrive at the array in the form of a plane wave. This means that there is a large distance between the transmitter and receiver (the receiver is in the far field of the transmitter). The array elements are separated by a distance d which must be less than or equal to half the wavelength (similar to the concept of minimum sampling frequency in DSP). Now we can see that the second ray travels an excess distance $d \cos \theta$. Similarly, the third and fourth rays travel an excess distance of $2d\cos\theta$ and $3d\cos\theta$, respectively. In array processing it is this excess distance between the arriving rays that is important, absolute distance from the source does not matter (unless you are interested in large-scale effects such as path loss). This excess distance between the different rays determines if the signals are going to add constructively or destructively (Fig. 6.1).

$$r = e^{-j\omega t}$$

$$r = e^{-j2\pi ft}$$

$$r = e^{-j2\pi\frac{c}{\lambda}t}$$

$$r = e^{-j2\pi\frac{x}{\lambda}}$$

$$r_{total} = \sum_{n=1}^{N_r} e^{-j2\pi\frac{x_n}{\lambda}}$$

Given below is the MATLAB code for the scenario shown in the figure above. We have considered two methods, one employing a "for-loop" and another using matrix manipulation. The second method is usually preferred as it is much faster and also allows us to directly apply techniques from linear estimation theory. We have plotted the array pattern for four cases with $N = 2, 4, 6, 8$. It is seen that as the number of array elements increases the gain (or directivity) of the array increases. In the case shown below we have considered that the four received signals are added with equal

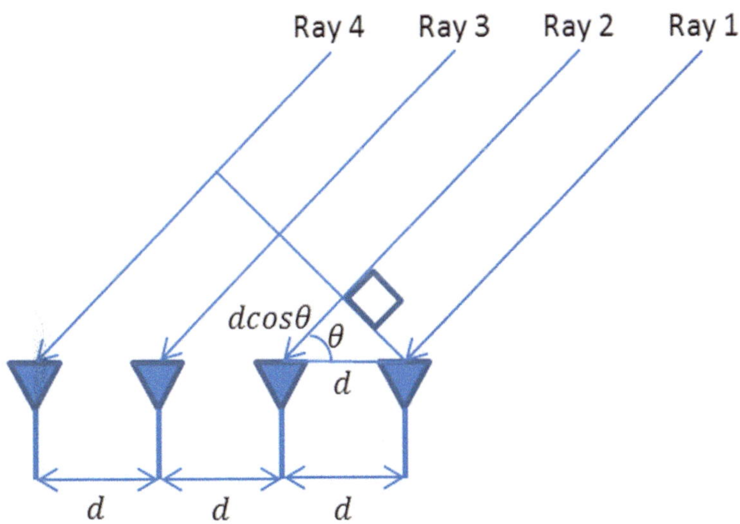

Fig. 6.1 Plane wave impinging upon a ULA (four elements)

weights ($w = 1$), but these weights can be adjusted to get various beam patterns (weights are typically complex quantities adjusting both phase and amplitude of the signal). This is typically called beamforming, and we will discuss this in the future post (Fig. 6.2).

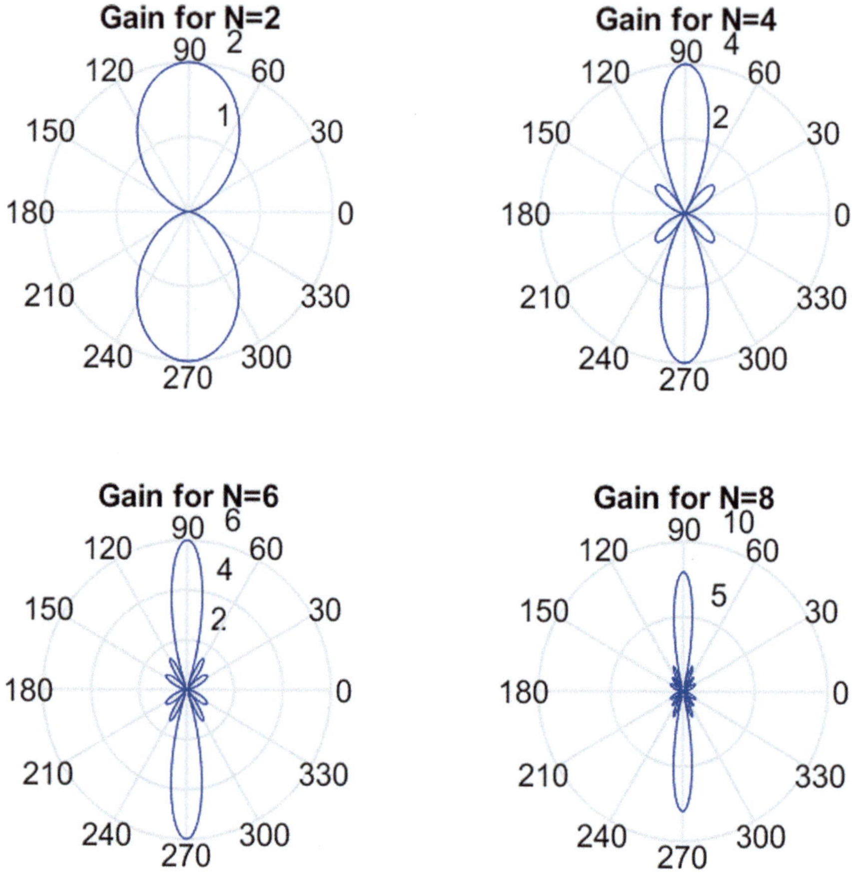

Fig. 6.2 Uniform linear array (ULA) with antenna length N equal to 2, 4, 6 and 8

```
%%%%%%%%%%%%%%%%%%%%%%%%%%%%%%%%%%%%%%%
%              SIMPLE UNIFORM LINEAR ARRRAY
%           WITH VARIABLE NUMBER OF ELEMENTS
%                f is the carrier frequency in Hz
%                c is the speed of light in m/sec
%                  l is the wavelength in m
%              d is the inter-element spacing in m
%                  Copyright 2020 RAYmaps
%%%%%%%%%%%%%%%%%%%%%%%%%%%%%%%%%%%%%%%
clear all
close all

f=1e9;
c=3e8;
l=c/f;
d=l/2;
no_elements=4;
theta=0:pi/180:2*pi;
r=zeros(1,length(theta));

for n=1:no_elements
  dx(n,:)=(n-1)*d*cos(theta);
  r=r+exp(-i*2*pi*(dx(n,:)/l));
end

polar(theta,abs(r),'b')
title ('Gain of a Uniform Linear Array')
%%%%%%%%%%%%%%%%%%%%%%%%%%%%%%%%%%%%%%%
```

```
%%%%%%%%%%%%%%%%%%%%%%%%%%%%%%%%%%%%%%%
%              SIMPLE UNIFORM LINEAR ARRRAY
%           WITH VARIABLE NUMBER OF ELEMENTS
%                MATRIX IMPLEMENTATION
%                f is the carrier frequency in Hz
%                c is the speed of light in m/sec
%                  l is the wavelength in m
%              d is the inter-element spacing in m
%                  Copyright 2020 RAYmaps
%%%%%%%%%%%%%%%%%%%%%%%%%%%%%%%%%%%%%%%
clear all
close all

f=1e9;
c=3e8;
l=c/f;
d=l/2;
no_elements=4;

theta=0:pi/180:2*pi;
n=1:no_elements;
n=transpose(n);
A=(n-1)*(i*2*pi*d*cos(theta)/l);
X=exp(-A);
w=ones(1,no_elements);
r=w*X;

polar(theta,abs(r),'r')
title ('Gain of a Uniform Linear Array')
%%%%%%%%%%%%%%%%%%%%%%%%%%%%%%%%%%%%%%%
```

Note:

For a uniform linear array with N elements and half wavelength interelement spacing, the Half Power Beam Width (HPBW) can be estimated as $1.78/N$ radians [1]. For the four-element case shown above the formula gave a HPBW of 25.49°, whereas our simulation yielded 26.20 degrees. For ten element case the formula gave a HPBW of 10.19°, whereas the simulation result was 10.20°. Similarly, the result for 20 elements is also quite accurate. So, we can say that the formula does help us to get a ballpark estimate and gives progressively more accurate results as the number of elements is increased. For a general case where the interelement spacing is not equal to half wavelength the formula is 0.89 (wavelength/total aperture length). Lastly, for those who still do not know what Half Power Beam Width also known as 3 dB bandwidth means, it is the width of the main lobe in degrees 3 dB down from the peak value of the radiation pattern.

6.3 Basics of Beamforming in Wireless Communications

In the previous section we had discussed the fundamentals of a uniform linear array (ULA). We had seen that as the number of array elements increases the gain or directivity of the array increases. We also discussed the Half Power Beam Width (HPBW) that can be approximated as $0.89(2/N)$ radians. This is quite an accurate estimate provided that the number of array elements N is sufficiently large (Fig. 6.3).

But the max gain is always in a direction perpendicular to the array (broadside). What if we want the array to have a high gain in another direction such as at 45°? How can we achieve this? This has application in radars where you want to search for

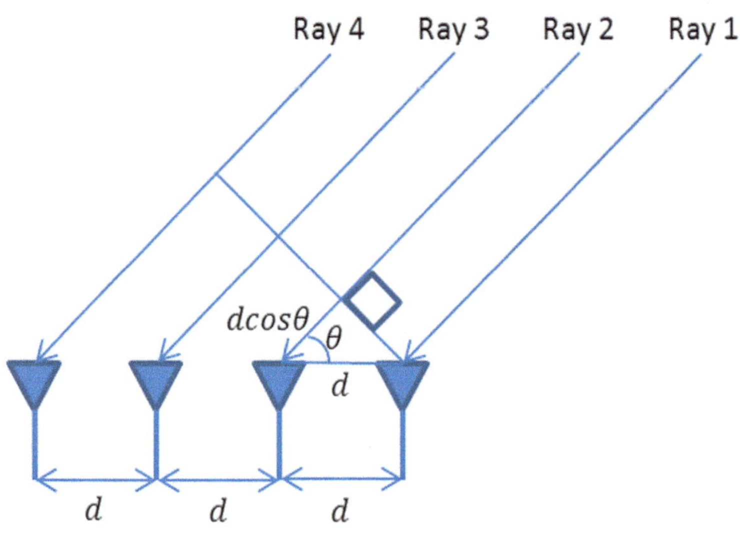

Fig. 6.3 Plane wave impinging upon a ULA

a target by scanning over 360° or in mobile communications where you want to send a signal to a particular user without causing interference to other users. One simple way is to physically rotate the antenna but that is not always a feasible solution.

Going back to the basics remember that the electric field pattern depends upon the constructive and destructive interference of incoming waves. If we have a vector (usually called the steering vector) that aligns the rays coming in from a particular direction, we would get a high gain in that direction. Similarly, we can steer a null in a particular direction if we want to reject a particular signal. This is something that needs further discussion.

The figure below shows the electric field pattern of a 10-element array steered toward 0, 30, 60 and 90°, respectively. We see that selectivity of the array is higher on the broadside than on the endfire. This has to do with how the cosine function behaves from 0 to 90°. The rate of change of cosine function is much faster around 90 degrees than at 0 degrees or 180°. The slowly changing cosine in the latter case causes a wide response on the endfire (Fig. 6.4).

```
%%%%%%%%%%%%%%%%%%%%%%%%%%%%%%%%%%%%%%%%%%%%%%%
%                    BEAMFORMING USING A
%                    UNIFORM LINEAR ARRRAY
%
%                    f is the frequency in Hz
%                    c is the speed of light in m/sec
%                    l is the wavelength
%                    d is the separation between array elements
%                    phi is the steering angle
%                    Copyright 2020 RAYmaps
%%%%%%%%%%%%%%%%%%%%%%%%%%%%%%%%%%%%%%%%%%%%%%%

clear all
close all

f=1e9;
c=3e8;
l=c/f;
d=l/2;
no_elements=10;
phi=pi/6;

theta=0:pi/180:2*pi;
n=1:no_elements;
n=transpose(n);

X=exp(-i*(n-1)*2*pi*d*cos(theta)/l);
w=exp( i*(n-1)*2*pi*d*cos(phi)/l);
w=transpose(w);
r=w*X;

polar(theta,abs(r),'b')
title ('Gain of a Uniform Linear Array')
%%%%%%%%%%%%%%%%%%%%%%%%%%%%%%%%%%%%%%%%%%%%%%%
```

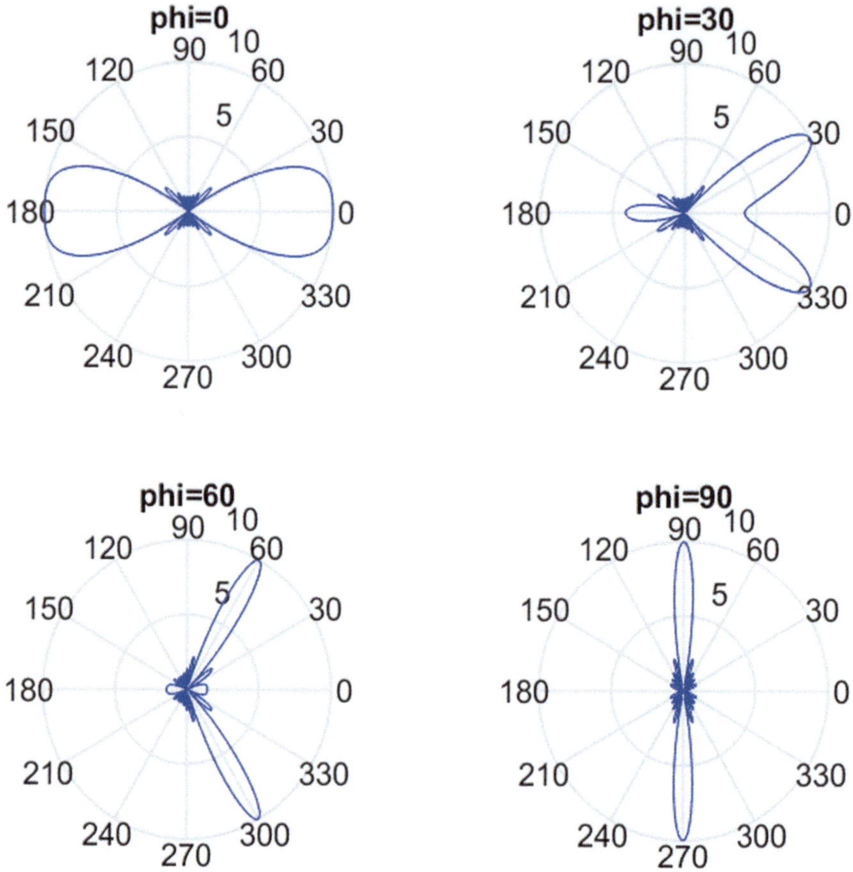

Fig. 6.4 Gain pattern for various steering angles

We did calculate the HPBW for a range of steering angles and found that it varied widely from as small as 10.20° to as large as 68.90°. This shows that simple beamforming using a steering vector has its limitations. The detailed results along with a graph are shown below (Fig. 6.5). It is seen that as the steering angle increases from about 5 degrees there is a sudden increase in HPBW, and then it suddenly drops by about 40° for one degree increase of steering angle from 24 to 25°. This is due to the fact that as the steering angle increases from zero degrees the main beam is split into two. Thus, the beam width calculated is of two beams, not a single beam as is usually the case. Beyond 25° we again lock on to a single beam and the HPBW is more realistic (Fig. 6.5).

Case 1: phi = 0, HPBW = 48.68°
Case 2: phi = 30, HPBW = 21.72°
Case 3: phi = 60, HPBW = 11.81°
Case 4: phi = 90, HPBW = 10.20°

For further visualization of the variation in antenna pattern as a function of the steering angle please have a look at this Interactive Graph (https://www.geogebra.org/m/ArF3sKpW). The parameters that can be varied include the angle of the beam,

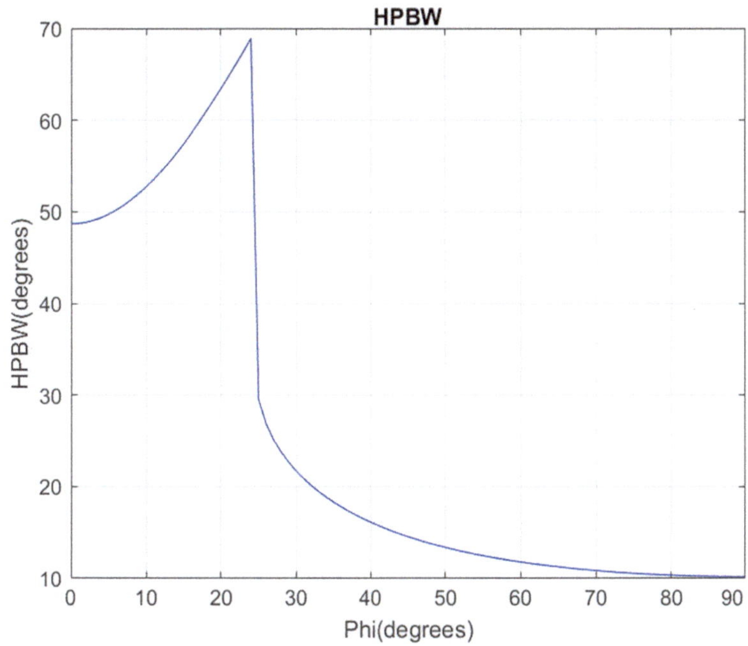

Fig. 6.5 Half power beam width of a ULA as a function of steering angle (Phi)

number of antenna elements and separation of the antenna elements. This is taken from an excellent online resource by the name of Geogebra. For further information on how you can use this tool for your own mathematical problems please do visit their website.

Note: A word of caution on how the HPBW is calculated as the steering angle phi is varied from 0 degrees to 90 degrees. As per the definition of HPBW, it is the angular width measured on the major lobe of an antenna radiation pattern at half power points. But what is the major lobe and how 3dB points are defined when there are multiple main beams with equal peak power? Additionally, should we limit our search to 0 to 180 degrees, -90 to 90 degrees, or scan the entire angular space? In our opinion, there is no right or wrong answer here, it depends upon the scenario being studied and our interpretation of it. Lastly, I would like to add that beam splitting can also generate misleading results and should be dealt with care.

6.4 Multicarrier Beamforming at MmWave

We have previously discussed beamforming for single carrier signals. Now we turn our attention to multicarrier signals particularly at mmWave where the bandwidths are two orders of magnitude (100×) higher than at sub 6 GHz band. We want to investigate that whether there is any distortion in the array response due to high signal bandwidths at mmWave.

But let us start with the case that we have discussed so far, i.e., 1 GHz single carrier case and a uniform linear array (ULA). We then add two other carriers at 1.2 GHz and 0.80 GHz, quite an extreme case, stretching the bandwidth to 400 MHz. Antenna spacing is still $\lambda/2 = 0.15$ m corresponding to the center frequency of 1 GHz.

So basically, we are breaking the spatial sampling theorem ($d < \lambda/2$) at 1.2 GHz where the element spacing should be less than 0.1250 m. We are still OK at 800 MHz where the element spacing must be less than 0.1875 m. But we must clarify that the sampling theorem is given for the worst case, when the signal is arriving from the endfire of a ULA. For a plane wave arriving from the broadside, all rays would add up coherently irrespective of interelement spacing.

Below we give the simulation results and MATLAB code used to generate these results. First, we consider the standard 1 GHz case (with 400 MHz bandwidth), and then we venture on to the more interesting case of a mmWave signal with three carriers at 37.0 GHz, 38.5 GHz and 40.0 GHz (total bandwidth of 3.0 GHz). These are not some random numbers; there is actually a mmWave frequency allocation at 37 GHz–40 GHz.

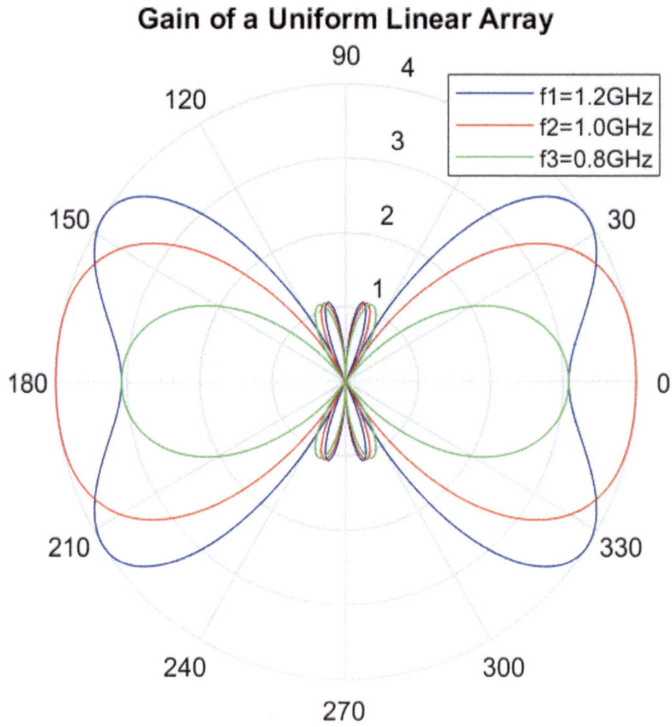

Fig. 6.6 ULA with four elements—endfire response at 1 GHz

It is seen that (Figs. 6.6 and 6.7) the departure of array response from the ideal depends upon both the bandwidth of the signal ($\Delta f = f_{\max} - f_{\min}$) as well as the center frequency (f_c). In fact, it depends on the ratio ($\Delta f/f_c$), and the smaller this ratio the better. As discussed above, at the lower frequency of 1 GHz this ratio is 0.40 ($\Delta f/f_c = 0.4/1.0$), whereas at mmWave this ratio is only 0.0779 ($\Delta f/f_c =$ 3.0/38.5). While there is some deviation in array response in the first case there is hardly any in the second.

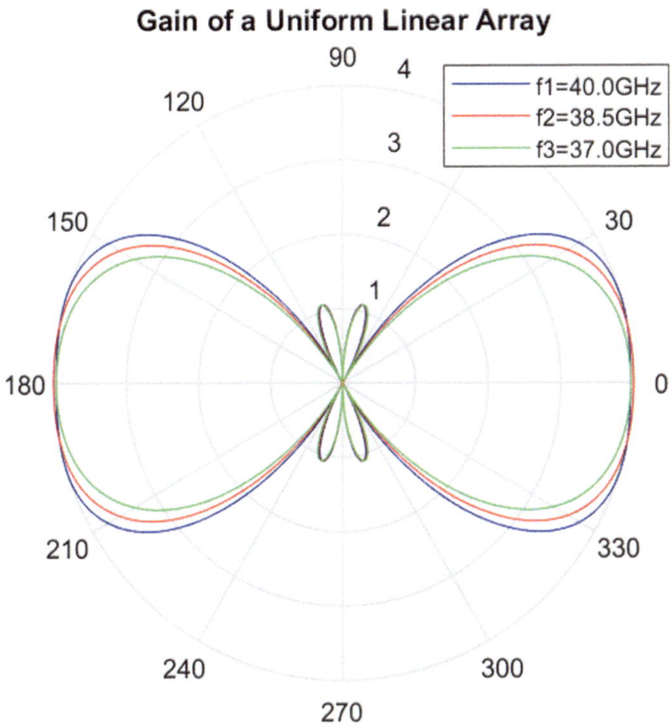

Fig. 6.7 ULA with four elements—endfire response at 38.5 GHz

```
%%%%%%%%%%%%%%%%%%%%%%%%%%%%%%%%%%%%%%%%
%                SIMPLE UNIFORM LINEAR ARRAY
%           WITH VARIABLE NUMBER OF ELEMENTS
%     f1, f2, f3 are the three frequencies of interest in Hz
%     l1, l2, l3 are the three wavelengths of interest in m
%              c is the speed of light in m/sec
%             d is the inter-element spacing in m
%                Copyright 2020 RAYmaps
%%%%%%%%%%%%%%%%%%%%%%%%%%%%%%%%%%%%%%%%

clear all
close all

f1=1.2e9;
f2=1.0e9;
f3=0.8e9;
c=3e8;

l1=c/f1;
l2=c/f2;
l3=c/f3;
d=l2/2;

no_elements=2;
n=1:no_elements;
n=transpose(n);

w=exp((n-1)*(i*2*pi*d*cos(pi/2)/l2));
w=transpose(w);
theta=0:pi/180:2*pi;
A=(n-1)*(i*2*pi*d*cos(theta)/l1);
X=exp(-A);
r=w*X;
polar(theta,abs(r),'b-')
hold on

A=(n-1)*(i*2*pi*d*cos(theta)/l2);
X=exp(-A);
r=w*X;
polar(theta,abs(r),'r-')
hold on

A=(n-1)*(i*2*pi*d*cos(theta)/l3);
X=exp(-A);
r=w*X;
polar(theta,abs(r),'g-')
hold off

title ('Gain of a Uniform Linear Array')
legend('f1=1.2GHz','f2=1.0GHz','f3=0.8GHz')
```

Note:

1. We can observe the appearance of grating lobes at 1.2 GHz. These lobes have the same magnitude as those at 1.0 GHz; however, they make the array response to be much wider along endfire.
2. We can easily rotate the main beam toward broadside by modifying the angle phi of the Steering Vector from 0 to pi/2 as given in the following piece of code: $w = \exp((n - 1) * (i * 2 * \text{pi} * d *\cos(\text{phi})/l2))$.
3. Here we have assumed that the response of the array to a wideband signal is equal to sum of the responses of the individual narrowband signals, so we can analyze their responses independently. But in reality, there are nonlinear effects such as intermodulation effects and we need to calculate the response of the composite signal.

Bottom Line

Due to the high carrier frequencies at mmWave the high bandwidth of signals has minimal impact, so traditional beamforming techniques would still work. However, one must be careful not to overlook nonlinear effects, and this would be the subject of a future post.

6.5 Rectangular Array—Mathematical Model and Code

In the previous few articles we discussed the fundamentals of uniform linear arrays (ULAs), beamforming and multicarrier beamforming. Now we turn our attention to more complicated array structures such as rectangular, triangular and circular. We still assume each element of the array to have an isotropic or omnidirectional (in the plane of the array) radiation pattern. The mathematical models for more complicated radiation patterns are an extension of what is developed here.

In this post we consider a square array which is a special case of rectangular array. We build up from the most basic case of a 2×2 array and derive the equation of the resultant signal, which is simply the summation of the individual signals received at the four array elements. Later on, we give the MATLAB code which can be used to plot the radiation pattern of any size rectangular array (Figs. 6.8 and 6.9).

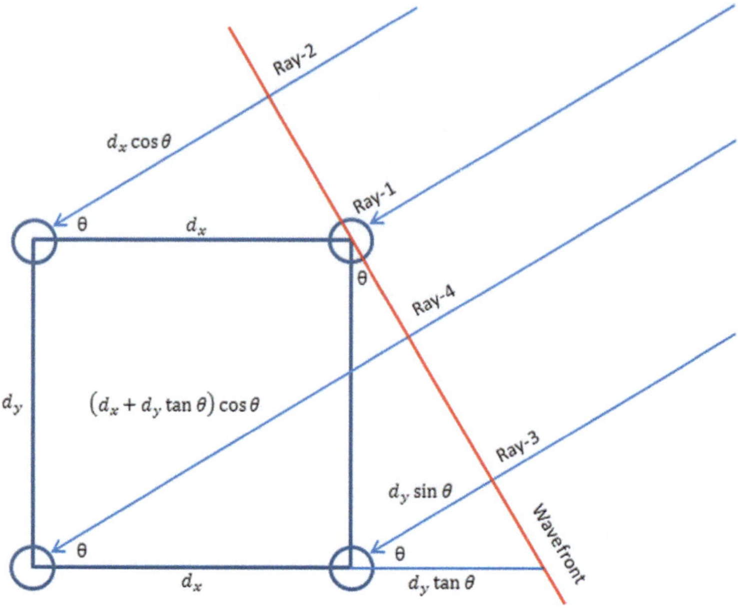

Fig. 6.8 Rectangular array geometrical representation

Let's consider a plane wave impinging upon a 2×2 receive array, the plane wave lies in the plane of the array. The source is considered to be in the far field of the receive array making the plane wave assumption to be realistic. The equations for the received signal at the four array elements are given below. Please note that since the combined signal only depends upon the relative phase of the four components, we assume the phase at the wavefront (red line in the figure below) to be zero. Also note that we have assumed that $d_x = d_y = d$.

$$
\begin{aligned}
r_1 &= 1 \\
r_2 &= e^{-j2\pi d cos\theta/\lambda} \\
r_3 &= e^{-j2\pi d sin\theta/\lambda} \\
r_4 &= e^{-j2\pi d(cos\theta + sin\theta)/\lambda} \\
r_t &= r_1 + r_2 + r_3 + r_4
\end{aligned}
$$

In general, for an $N \times M$ array the resultant signal can be written as

$$
r_t = \sum_{n=0}^{N-1} \sum_{m=0}^{M-1} e^{-j2\pi d(n\cos\theta + m\sin\theta)/\lambda}
$$

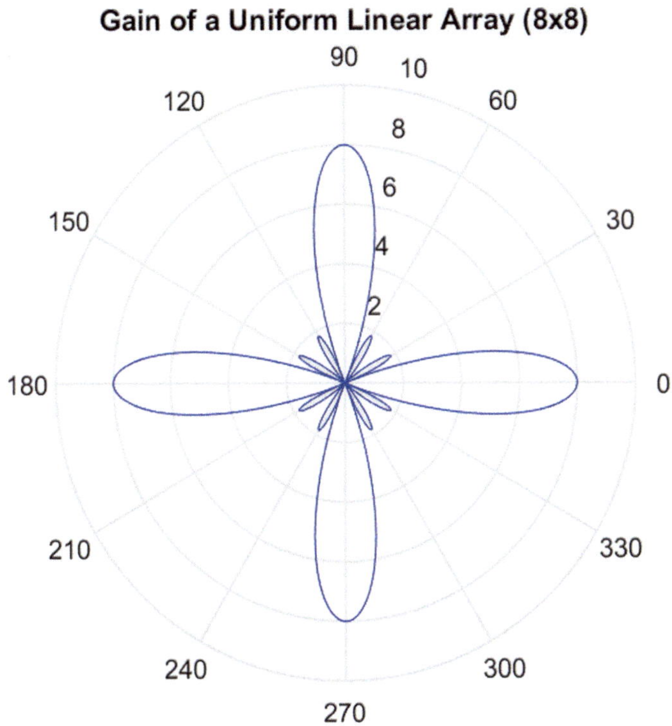

Fig. 6.9 Square array radiation pattern in the plane of the array

When the separation along the x and y axes is not the same, we have

$$r_t = \sum_{n=0}^{N-1} \sum_{m=0}^{M-1} e^{-j2\pi\left(nd_x\cos\theta + md_y\sin\theta\right)/\lambda}$$

The range of n and m is 0 to $N - 1$ and 0 to $M - 1$, respectively.

```
%%%%%%%%%%%%%%%%%%%%%%%%%%%%%%%%%%%%%%%%%%%%%%%%
%                    SIMPLE RECTANGULAR ARRRAY
%                       WITH N x M ELEMENTS
%                    f is the carrier frequency in Hz
%                    c is the speed of light in m/sec
%                       l is the wavelength in m
% dx and dy are the inter-element spacings along the two axes
%     NxM is the number of elements in the receive array
%                      Copyright 2020 RAYmaps
%%%%%%%%%%%%%%%%%%%%%%%%%%%%%%%%%%%%%%%%%%%%%%%%
clear all
close all

f=1e9;
c=3e8;
l=c/f;
dx=l/3;
dy=l/3;
theta=0:pi/1800:2*pi;
N=8;
M=8;

r=0;
for n=1:N
  for m=1:M
    r=r+exp(-i*2*pi*(dx*(n-1)*cos(theta)+dy*(m-1)*sin(theta))/l);
  end
end

polar(theta,abs(r),'bo-')
title ('Gain of a Uniform Linear Array')
%%%%%%%%%%%%%%%%%%%%%%%%%%%%%%%%%%%%%%%%%%%%%%%%
```

In the above code we have assumed a carrier frequency of 1 GHz which gives us a wavelength of 0.3 m. The element separation along the x and y axis is assumed to be 0.1 m (lambda/3). The total number of elements is $8 \times 8 = 64$. The maximum gain obtained is equal to 8 on the linear scale or 9 dB on the logarithmic scale (not shown here but the gain perpendicular to the array would be 18 dB). The radiation pattern has four peaks which does not make this array structure to be of much practical use. More complex array structures resulting in more desirable radiation patterns will be discussed in the future sections.

Note:

Please note that we have simplified the problem significantly by assuming an omnidirectional pattern of the array elements and plotting the composite radiation pattern only in the plane of the array. In reality the array elements do not always have an omnidirectional radiation pattern (one popular antenna that does have an omnidirectional pattern is a dipole), and we have to plot the 3D pattern of the array (or cuts along different planes) to get a better understanding of the characteristics such as Half Power Beam Width, First Null Beam Width and Sidelobe Level.

6.6 Circular Array—Mathematical Model and Code

In the previous section we discussed the case of a square array which is a special case of a rectangular array. The code we shared can handle both the cases as well as uniform linear array. We did briefly talk about the response of an element vs the response of an array, but we did not put forward the mathematical relationship. So here it is

$$\text{Response of an Array} = \text{Array Factor} \times \text{Element Factor}$$

In this post as well as previous posts we have assumed the element response to be isotropic (or at least omnidirectional in the plane of the array) giving us an Element Factor of 1. So, the array response is nothing but equal to the array factor. In this post we mostly discuss the 2D array factor but briefly touch upon the 3D case as well at the end.

As discussed in the previous posts when a plane wave impinges upon an array its absolute phase is not that important (although it might be important in synchronization at the receiver but we defer that discussion for the moment). What is important is the relative phase at the array elements. This can be calculated by first determining the excess path length at each element from a reference element and then adding up the contribution of each element to the array pattern.

The excess path length calculation, for the elements of a circular array, is not as straightforward as that for a uniform linear array or a Rectangular Array. We show two methods for calculation of the excess path length; you can choose whichever you prefer. It's no surprise that both methods give identical results. One point that needs to be clarified about the mathematical model below is that we drop the term $\cos\phi$ from the final equation. This is because it is a common term for all the array elements and does not have any impact on the composite pattern (array factor) (Figs. 6.10, 6.11 and 6.12).

Step 1 $\varphi_n + \theta_n + \theta_n = 180$ $2\theta_n = 180 - \varphi_n$ $\theta_n = 90 - \frac{\varphi_n}{2}$ **Step 2** $\theta_n + \psi_n + \varphi = 180$ $\psi_n = 180 - \varphi - \theta_n$ $\psi_n = 180 - \varphi - \left(90 - \frac{\varphi_n}{2}\right)$ $\psi_n = 90 - \varphi + \frac{\varphi_n}{2}$	**Step 3** $\frac{d_n}{2} = a \sin\frac{\varphi_n}{2}$ $d_n = 2a \sin\frac{\varphi_n}{2}$ **Step 4** $\delta_n = d_n \cos\psi_n$ $\delta_n = 2a \sin\left(\frac{\varphi_n}{2}\right)\cos\left(90 - \varphi + \frac{\varphi_n}{2}\right)$ $\delta_n = 2a \sin\left(\frac{\varphi_n}{2}\right)\sin\left(\varphi - \frac{\varphi_n}{2}\right)$ $\delta_n = a[\cos(\varphi - \varphi_n) - \cos\varphi]$
$AF = \sum_{n=0}^{N-1} e^{-jka\cos(\varphi-\varphi_n)}$	

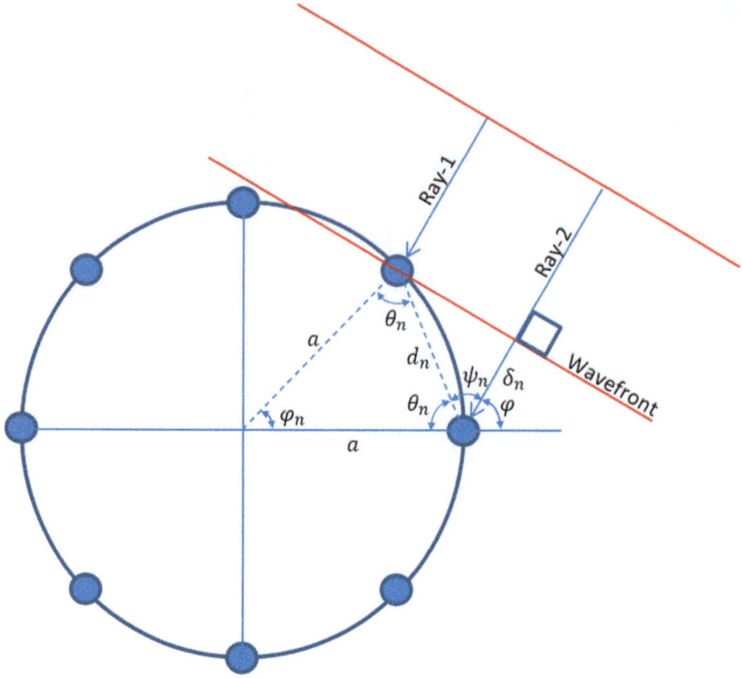

Fig. 6.10 Circular array derivation using method 1

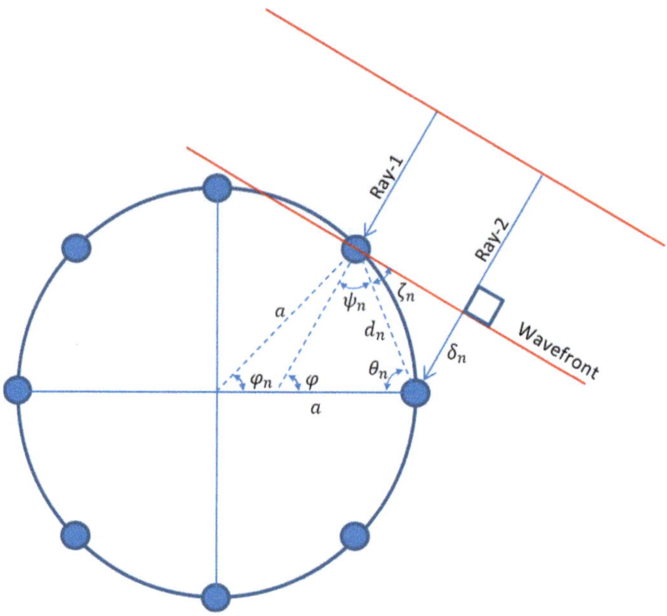

Fig. 6.11 Circular array derivation using method 2

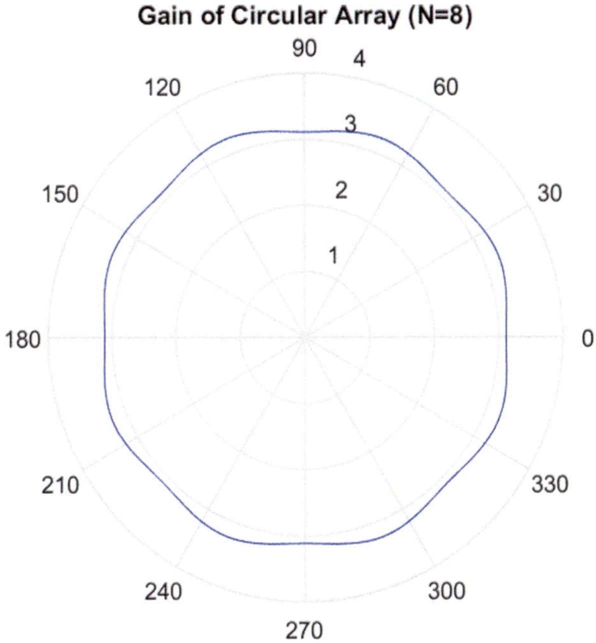

Fig. 6.12 Eight element circular array

Step 1 $\varphi_n + \theta_n + \theta_n = 180$ $2\theta_n = 180 - \varphi_n$ $\theta_n = 90 - \frac{\varphi_n}{2}$ **Step 2** $\psi_n + \theta_n + \varphi = 180$ $\psi_n = 180 - \varphi - \theta_n$ $\psi_n = 180 - \varphi - \left(90 - \frac{\varphi_n}{2}\right)$ $\psi_n = 90 - \varphi + \frac{\varphi_n}{2}$	**Step 3** $\psi_n + \zeta_n = 90$ $\zeta_n = 90 - \psi_n$ $\zeta_n = 90 - \left(90 - \varphi + \frac{\varphi_n}{2}\right)$ $\zeta_n = \varphi - \frac{\varphi_n}{2}$ **Step 4** $\delta_n = d_n \sin \zeta_n$ $\delta_n = 2a \sin\left(\frac{\varphi_n}{2}\right) \sin\left(\varphi - \frac{\varphi_n}{2}\right)$ $\delta_n = a[\cos(\varphi - \varphi_n) - \cos \varphi]$
$AF = \sum_{n=0}^{N-1} e^{-jka\cos(\varphi - \varphi_n)}$	

```
%%%%%%%%%%%%%%%%%%%%%%%%%%%%%%%%%%%%%%%%
%           CIRCULAR ARRAY WITH VARIABLE
%         NUMBER OF ELEMENTS AND SPACING
%              f is the carrier frequency in Hz
%              c is the speed of light in m/sec
%                I is the wavelength in m
%            d is the inter-element spacing in m
%       N is the number of elements in the receive array
%                Copyright 2020 RAYmaps
%%%%%%%%%%%%%%%%%%%%%%%%%%%%%%%%%%%%%%%%
clear all
close all

f=1e9;
c=3e8;
l=c/f;
k=(2*pi)/l;
N=8;
n=0:N-1;
phi_n=2*pi*n/N;
phi=0:pi/180:2*pi;
M=length(phi);
d_circular=l/2;
a=(N*d_circular)/(2*pi);

for m=1:M
  AF(m)=sum(exp(-i*k*a*(cos(phi(m)-phi_n))));
end

polar(phi, abs(AF))
%%%%%%%%%%%%%%%%%%%%%%%%%%%%%%%%%%%%%%%%
```

It is seen that array pattern for 8-element array has almost a uniform response with a gain of about 3. Here the element separation, along the circumference of the ring, is set to half the wavelength. Different antenna patterns can be generated by varying the interelement spacing (which determines the radius of the ring). More complicated patterns can be generated by using multiring antenna formations [2]. Also, it must be noted that we are simply adding the signal arriving at each of the elements without altering the phase or the amplitude. In actual implementations there is always a weighting pattern applied to the signals arriving at the array elements to get more useful patterns [3].

Lastly, I would like to briefly comment on the 3D patterns that can be generated by using 3D array factors, a natural extension to the models discussed above. The equation of the 3D array factor is really not that complicated and has just one additional variable θ, which is nothing but the angle of elevation. In the above we have assumed that the angle of elevation to be 90°. If the angle of elevation is not 90°, that is we are not in the plane of the array, the array factor is modified to

$$\sum_{n=0}^{N-1} e^{-jka(\cos\varphi\cos\varphi_n\sin\theta+\sin\varphi\sin\varphi_n\sin\theta)}$$

$$\sum_{n=0}^{N-1} e^{-jka\sin\theta(\cos\varphi\cos\varphi_n+\sin\varphi\sin\varphi_n)}$$

$$\sum_{n=0}^{N-1} e^{-jka\sin\theta(\cos(\varphi-\varphi_n))}$$

```
%%%%%%%%%%%%%%%%%%%%%%%%%%%%%%%%%%%%
%  3D PATTERN OF CIRCULAR ARRAY WITH VARIABLE
%         NUMBER OF ELEMENTS AND SPACING
%              f is the carrier frequency in Hz
%              c is the speed of light in m/sec
%                l is the wavelength in m
%          d is the inter-element spacing in m
%     N is the number of elements in the receive array
%               Copyright 2020 RAYmaps
%%%%%%%%%%%%%%%%%%%%%%%%%%%%%%%%%%%%
clear all
close all

f=1e9;
c=3e8;
l=c/f;
k=(2*pi)/l;
N=8;
n=0:N-1;
phi_n=2*pi*n/N;
phi=-pi:pi/10:pi;
theta=0:pi/36:pi/2;
M=length(theta);
P=length(phi);
d_circular=l/2;
a=(N*d_circular)/(2*pi);

for p=1:P
 for m=1:M
  AF(m,p)=sum(exp(-i*k*a*sin(theta(m))*(cos(phi(p)-phi_n))));
  x(m,p)=abs(AF(m,p))*sin(theta(m))*cos(phi(p));
  y(m,p)=abs(AF(m,p))*sin(theta(m))*sin(phi(p));
  z(m,p)=abs(AF(m,p))*cos(theta(m));
 end
end

mesh(x,y,z)
```

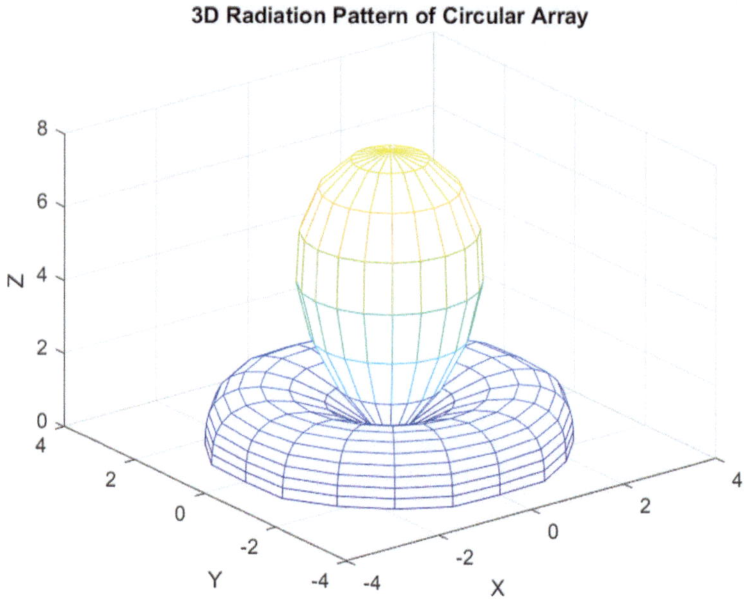

Fig. 6.13 3D pattern of eight element circular array

If we fix the value of ϕ to be zero and vary θ from -90° to $+90^\circ$ we get a very interesting pattern (in the figure above we vary theta from 0 to 90 degrees and phi from –180 degrees to 180 degrees). This pattern shows that with isotropic array elements it is possible to have a maximum gain of 8 on the linear scale (or 9 dB on the logarithmic scale), which is expected from an array with 8 elements and Element Factor of unity. This can also be verified from the above equation by realizing that $\sin\theta = \sin(0) = 0$ and anything raised to the power of zero is one. Eight of these terms added together give a total gain of 8 (Fig. 6.13).

Note:

Interelement separation on the circumference is labeled as "d_circular," and this is set to half the wavelength [3]. Based upon this the radius of the ring is calculated.

6.7 Direction of Arrival Estimation

Direction of Arrival (DOA) estimation is a fundamental problem in communications and signal processing with application in cellular communications, radar, sonar, etc. It has become increasingly important in recent times as 5G communications use DOA to spatially separate the users resulting in higher capacity and throughput. Direction of Arrival estimation can be thought of as the converse of beamforming. As you might recall from the discussion in previous sections, in beamforming you use the steering vector to receive a signal from a particular direction, rejecting the signals from other directions. In DOA estimation you scan the entire angular domain to find the required signal or signals and estimate their angles of arrival and possibly the ranges as well.

The theory of DOA estimation draws from the techniques and algorithms used for spectral estimation, as finding the angle of arrival in the spatial domain is the same as finding the frequency of a windowed signal in the temporal domain. Just as we have to follow the Nyquist's sampling theorem in time domain, we have to follow the spatial sampling theorem when implementing beamforming or DOA estimation. Some of the popular techniques used for DOA estimation are correlation, MVDR, FFT, MUSIC and ESPRIT [4–6]. We only discuss the first two in this section as the remaining ones are discussed in detail in the chapter on phase and frequency.

Correlation or Delay and Sum (DS) method, as it is sometimes referred to, is the simplest and most effective technique and works quite well at low SNRs and with moderate array size. Minimum Variance Distortionless Response (MVDR) on the other hand is a bit more complicated as it requires estimation of the correlation matrix and also requires couple of matrix inversions. As you may already know matrix inversion is a tricky subject and if the matrix is ill conditioned the inverse may not even exist. A technique usually used to get around this problem is to use the Moore Penrose Pseudo Inverse instead of direct inversion. Another problem with MVDR is that for estimating the correlation matrix you need multiple snapshots to get good results.

Given below is the code and simulation results (Figs. 6.14 and 6.15) for both the techniques. We have assumed a frequency of 1 GHz, antenna spacing of $\lambda/2$, angles of arrival of 30, 60 and 90 degrees, signal to noise ratio of 0 dB (per element) and scan resolution of 1 degree over azimuth. Antenna array size is varied from 20 to 100 elements, and correlation matrix for MVDR is calculated using 1000 snapshots (no averaging is used for DS). It is seen that performance of simple correlation scheme is quite respectable even for a receive array size of 20 and improves appreciably when the array size is increased to 100 (pencil beam is obtained). This is not the case for MVDR which is highly sensitive to noise level and performance is not that great even for $N = 100$. However, a surprising result was that MVDR performs better than DS in very poor SNR (which can be attributed to averaging used in calculation of correlation matrix for MVDR). Mathematically the two methods can be described as:

$$P_{DS} = E\left|w^H y\right|^2 = w^H R_{yy} w$$

$$P_{MVDR} = \frac{1}{w^H R_{yy}^{-1} w}$$

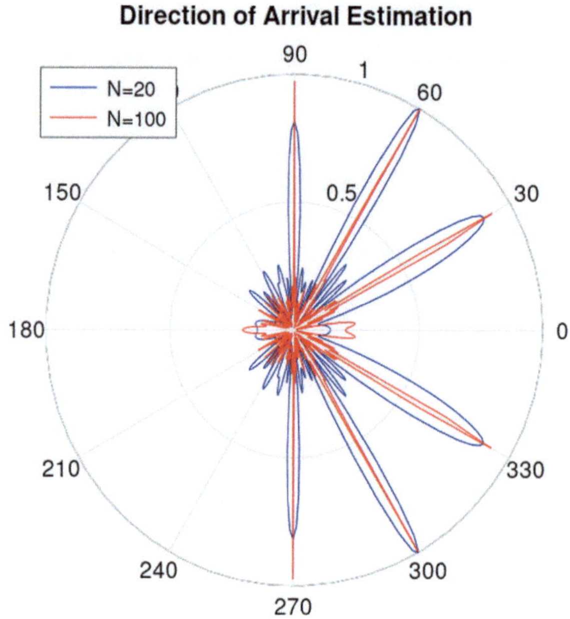

Fig. 6.14 Correlation-based DOA estimation at SNR of 0 dB

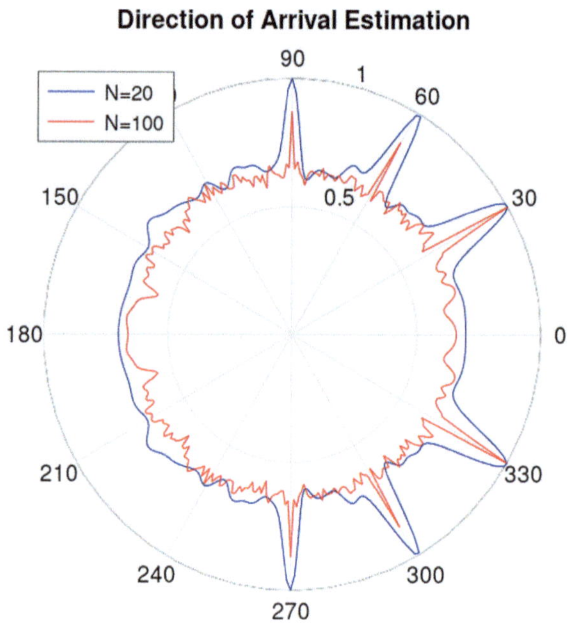

Fig. 6.15 MVDR-based DOA estimation at SNR of 0 dB

```
%%%%%%%%%%%%%%%%%%%%%%%%%%%%%%%%%%%%%%%%%%%%%%%%%%%%%
%              DIRECTION OF ARRIVAL ESTIMATION USING A ULA
%                          CORRELATION METHOD
%
%                       N is the number of elements
%                       d is the inter-element spacing
%                         theta is the angle of arrival
%
%                          Copyright 2022 RAYmaps
%%%%%%%%%%%%%%%%%%%%%%%%%%%%%%%%%%%%%%%%%%%%%%%%%%%%%
clear all
close all

f=1e9;
c=3e8;
l=c/f;
d=l/2;
N=20;
phi=0:pi/180:2*pi;


theta1=pi/6;
theta2=pi/3;
theta3=pi/2;
n=1:N;
n=transpose(n);

x1=exp(-i*(n-1)*2*pi*d*cos(theta1)/l);
x2=exp(-i*(n-1)*2*pi*d*cos(theta2)/l);
x3=exp(-i*(n-1)*2*pi*d*cos(theta3)/l);

x=x1+x2+x3;
y=x+0.7071*(randn(N,1)+i*randn(N,1));
w=exp(-i*(n-1)*2*pi*d*cos(phi)/l);
r=w'*y;

polar(phi,abs(r)/max(abs(r)),'b')
title ('Direction of Arrival Estimation')
%%%%%%%%%%%%%%%%%%%%%%%%%%%%%%%%%%%%%%%%%%%%%%%%%%%%%
```

```
%%%%%%%%%%%%%%%%%%%%%%%%%%%%%%%%%%%%%%%%%%%%%%%%%%
%               DIRECTION OF ARRIVAL ESTIMATION USING A ULA
%                              MVDR METHOD
%
%                        N is the number of elements
%                        d is the inter-element spacing
%                         theta is the angle of arrival
%
%                          Copyright 2022 RAYmaps
%%%%%%%%%%%%%%%%%%%%%%%%%%%%%%%%%%%%%%%%%%%%%%%%%%
clear all
close all

f=1e9;
c=3e8;
l=c/f;
d=l/2;
N=20;
phi=0:pi/180:2*pi;
theta=[pi/6, pi/3, pi/2];
n=transpose(1:N);

x1=exp(-i*(n-1)*2*pi*d*cos(theta(1))/l);
x2=exp(-i*(n-1)*2*pi*d*cos(theta(2))/l);
x3=exp(-i*(n-1)*2*pi*d*cos(theta(3))/l);
x=x1+x2+x3;

R=zeros(N,N);
for m=1:1000
   y=x+0.7071*(randn(N,1)+i*randn(N,1));
   R=R+y*y';
end
R=R/m;

R1=pinv(R,0.1);
w=exp(-i*(n-1)*2*pi*d*cos(phi)/l);
for k=1:length(phi);
   w1=w(:,k);
   P(k)=abs(pinv(w1'*R1*w1,0.1));
end

polar(phi,P/max(P),'b')
title ('Direction of Arrival Estimation')
%%%%%%%%%%%%%%%%%%%%%%%%%%%%%%%%%%%%%%%%%%%%%%%%%%
```

Note:

1. Please note that so far our analysis is based on visual inspection, which can be misleading. It is possible that a particular technique may have smaller peaks but

accuracy of angular information is greater. For more in depth analysis we need to compare the Mean Square Error (MSE) to Cramer Rao Lower Bound (CRLB). We also need to decrease the step size for the scan.
2. Please note that we have scanned the entire angular domain from 0 to 360° to find the DOA. In ULA theory it is known that response is symmetrical about the axis of the array so typically only 0 to 180° is scanned.
3. Some interesting antenna patterns can be obtained by using non-uniform antenna arrays and we hope to discuss this in the future.
4. Beamforming and DOA go hand in hand. DOA finds the angle of arrival and then the base station can form a beam on that angle using beamforming techniques.
5. Minimum Variance Distortionless Response is also known Capon estimator after its founder J. Capon who proposed his spectral analysis method in 1969.

6.8 Fundamentals of Linear Array Processing

We had previously discussed the fundamentals of array processing particularly the concepts of gain, directivity, array factor, beamforming and DOA. Now we build upon these concepts to introduce some linear estimation techniques that are used in array processing. These are particularly suited to a situation where multiple users are spatially distributed in a cell and they need to be separated based upon their angles of arrival. But first let us introduce the linear model; I am sure you have seen this before.

$$x = Hs + w$$

Here, s is the vector of symbols transmitted by M users, H is the $N \times M$ channel matrix, w is the noise vector of length N and x is the observation vector of length N. The channel matrix formed by the channel coefficients is deterministic (as opposed to probabilistic) in nature as it is purely dependent upon the phase shifts that the channel introduces due to varying path lengths between the transmit and receive antennas. The impact of a channel coefficient can be thought of as a rotation of the complex signal without altering its amplitude.

This means that the channel acts like a single tap filter, and the process of convolution is reduced to simple multiplication (a reasonable assumption if the symbol length is much larger than the channel delay spread). The channel model does not accommodate for path loss and fading that are also inherent characteristics of the channel. But the techniques are general enough for these effects to be factored in later. Furthermore, it is assumed that the channel H is known at the receiver. This is a realistic assumption if the channel is slowly varying and can be estimated by sending pilot signals (Fig. 6.16).

Going back to the linear model we see that we know x and H while s and w are unknown. Here w cannot be estimated since it's random in nature (remember what the term AWGN stands for?), and we ignore it for the moment. The structure of s

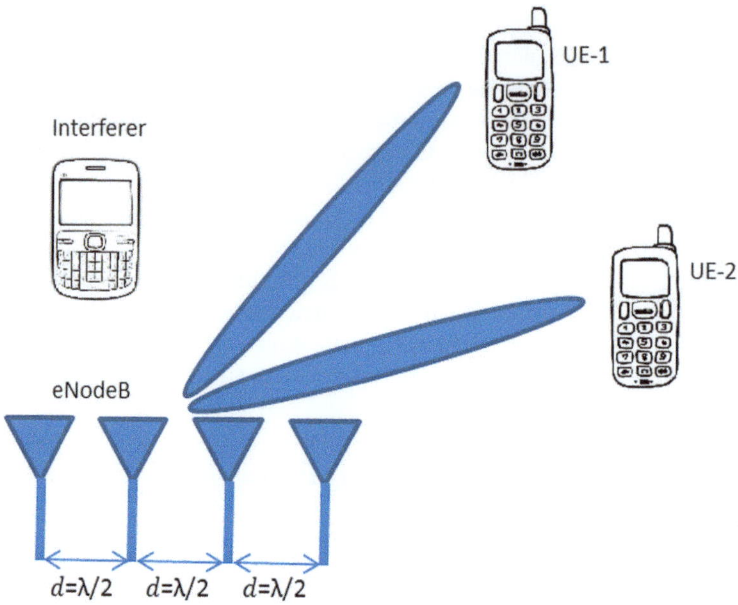

Fig. 6.16 Beamforming using a uniform linear array

is known. For example, if we are using BPSK modulation, then the M symbols of the signal vector s can either be $+\,1$ or -1. So, we can start the process of symbol detection by substituting all possible combinations of $s_1 s_2 \ldots s_M$ and determine the combination that minimizes

$$\|x - Hs\|$$

This is called the maximum likelihood (ML) solution as it determines the combination that was most likely to have been transmitted based upon the observation. Although ML is conceptually very appealing and yields good results, it becomes prohibitively complex as the constellation size or number of transmit antennas increases. For example, for 2-Transmit case and BPSK modulation there are 2^(1 bit $\times$ 2 antennas) = 2^2 = 4 combinations, which seems quite simplistic. But if 16-QAM modulation is used and there are 4-Transmit antennas, the number of combinations increases to 2^(4 bits $\times$ 4 antennas) = 2^16 = 65,536. So, we conclude that ML is not the solution we are looking for, if computational complexity is an issue (which might become less of an issue as the processing power of devices increases).

Next, we turn our attention to a technique popularly known as Zero Forcing or ZF. According to this technique the channel has a multiplicative effect on the signal. So, to remove this effect we simply divide the signal by the channel or in the language of matrices we perform matrix inversion. Mathematically we have

$$x = Hs + w$$

$$H^{-1}x = H^{-1}Hs + H^{-1}w$$

$$H^{-1}x = s + H^{-1}w$$

So, we see that we get back the signal s but we also get a noise component enhanced by inverse of the channel matrix. This is the well-known problem of ZF called noise enhancement. Then there are other problems such as non-existence of the inverse when the channel H is not a square matrix (the matrix is only square when the number of transmit and receive antennas is the same). The inverse of H also cannot be calculated if H is not full rank or determinant of H is zero. So, we now introduce another technique called Least Squares (LS). According to this the signal vector can be estimated as

$$\hat{s} = \left(H^H H\right)^{-1} H^H x$$

This is also sometimes referred to as the Minimum Variance Unbiased Estimator [7]. This can be easily implemented in MATLAB using Moore Penrose Pseudo Inverse or pinv(H). This is much more stable than going for the direct inversion methods. We next plot the bit error rate (BER) using the code below. The number of receive antennas is varied from two to ten while the number of transmit antennas is fixed at four. The transmit antennas are assumed to be positioned at 30, 40, 50 and 60° from the axis of the receive array (a better model is Uniform Random Line of Sight or UR-LoS but as expected it results in much poorer BER performance). The receive antennas are separated by $\lambda/2$ meters. The frequency of operation is 1 GHz, but it is quite irrelevant to the scenario considered as everything is measured in multiples of wavelengths. The E_b/N_o ratio is varied from 5 to 20 dB in steps of 5 dB (Fig. 6.17).

As expected, the BER for the two methods, other than ML (pseudoinverse and MATLAB implementation of pseudoinverse), is more or less the same and decreases rapidly once the number of receive antennas becomes greater than number of transmit antennas (or number of signals). The case where the number of receive antennas is less than number of signals (equal powered and with a small angular separation) is dealt with by Overloaded Array Processing (OLAP) techniques and have been discussed in detail by James Hicks [8].

Strangely enough it is seen that the overloaded case is not the worst part of the BER curve. The worst BER is observed when the number of transmit and receive antennas is the same (four in this case). In other words, the BER gradually increases as the

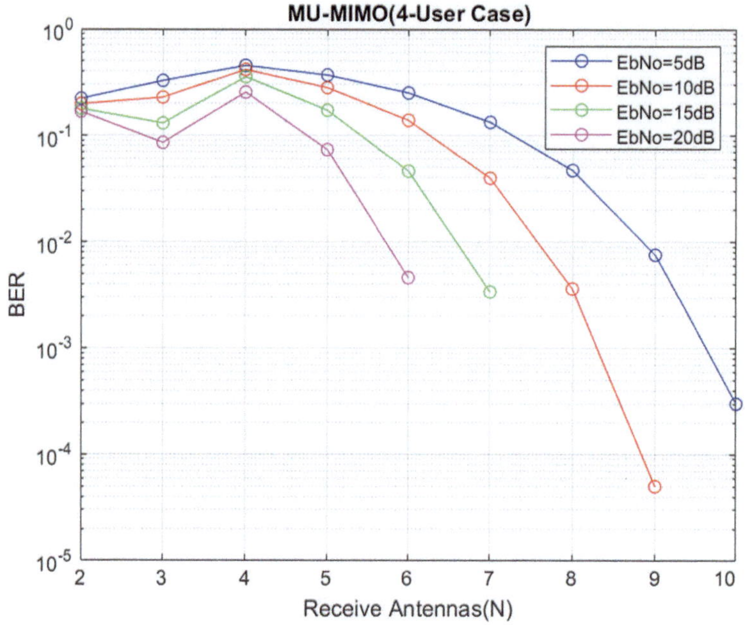

Fig. 6.17 Bit error rate for changing Rx array length

rank of the channel matrix increases and then decreases once it reaches its maximum value. This is quite an interesting result and needs to be further investigated. We also experimented with the MATLAB function pinv by changing the tolerance parameter (tol).

tol = max (size (H)) * sigma_max (H) * eps

where sigma_max(H) is the maximal singular value of channel matrix H and eps is the machine precision.

More precisely, eps is the relative spacing between any two adjacent numbers in the machine's floating point system. This number is obviously system dependent. On machines that support IEEE floating point arithmetic, eps is approximately 2.2204e-16 for double precision and 1.1921e-07 for single precision.

```
%%%%%%%%%%%%%%%%%%%%%%%%%%%%%%%%%%%%%
%               MULTIUSER DETECTION USING
%               A UNIFORM LINEAR ARRRAY
%
%               f is the carrier frequency in Hz
%               c is the speed of light in m/sec
%               l is the wavelength in m
%             d is the inter-element spacing in m
%     N is the number of elements in the receive array
%      M is the number of transmit antennas or users
%                Copyright 2020 RAYmaps
%%%%%%%%%%%%%%%%%%%%%%%%%%%%%%%%%%%%%

clear all
close all

% SETTING THE PARAMETERS FOR THE SIMULATION
f=1e9;
c=3e8;
l=c/f;
d=l/2;
N=10;

theta=([30 40 50 60])*pi/180;
EbNo=10;
sigma=1/sqrt(2*EbNo);

n=1:N;
n=transpose(n);
M=length(theta);

% RECEIVE SIGNAL MODEL (LINEAR)
s=2*(round(rand(M,1))-0.5);
H=exp(-i*(n-1)*2*pi*d*cos(theta)/l);
wn=sigma*(randn(N,1)+i*randn(N,1));
x=H*s+wn;

% PINV without tol
% y=pinv(H)*x;

% PINV with tol
y=pinv(H, 0.1)*x;

% DEMODULATION AND BER CALCULATION
s_est=sign(real(y));
ber=sum(s!=s_est)/length(s);
```

Note:

1. The maximum number of its linearly independent columns (or rows) of a matrix is called the rank of a matrix. The rank of a matrix cannot exceed the number of its rows or columns (lesser of the two). If we consider a square matrix, the columns (or rows) are linearly independent only if the matrix is non-singular.

2. We had to do a bit of trial and error to find the value of tol that gives the lowest bit error rate. The reader is encouraged to experiment with the code and find the value that is best suited to the scenario under study.

Questions and Numerical Problems

1. What is a perfect isotropic radiator? Give examples (if any). What is an omnidirectional radiator? Give examples (if any).
2. Define the E-field and H-field of a dipole antenna in the (a) near field and (b) far field. What is characteristic impedance?
3. What is the difference between antenna gain and directivity, explain. Give the expression for antenna gain of a (a) Monopole antenna (b) Dipole antenna.
4. What is antenna aperture, give the mathematical expression defining the antenna aperture. What is the significance of its relationship with wavelength?
5. What is the difference between array factor and Element Factor, how can they be used to calculate the radiation pattern of an array? Give an example.
6. Experiment with interelement spacing of a uniform linear array, particularly observe what happens when interelement spacing is reduced to less than half the wavelength.
7. Give examples of non-uniform linear arrays, in which scenarios can they be useful? Modify the code for uniform linear array to calculate and plot the array pattern for a non-uniform linear array.
8. What is antenna correlation, how is it calculated? What happens to the capacity of a MIMO system in presence of high antenna correlation?
9. Explain the relationship between beamforming and Direction of Arrival estimation using a uniform linear array.
10. What is Half Power Beam Width (HPBW), how is it calculated? Give MATLAB code for its calculation. Observe the behavior as the number of elements is varied.
11. In beamforming why is the width of the beam the narrowest at broadside, consider the case of a uniform linear array.
12. Using the mathematical relationships and code for a circular array simulate the case of concentric arrays, experiment with the number of elements and the circles so formed.
13. Just like we have plotted the 3D radiation pattern for a circular array, plot the 3D radiation pattern of a square or rectangular array. Observe the behavior.
14. Estimate the Direction of Arrival using ESPRIT and MUSIC algorithms (you might have to look at the next chapter on frequency estimation).
15. Modify the code for linear array processing, use the Uniform Random Line of Sight (UR-LOS) channel model in the simulation. Observe the behavior. Is the bit error rate performance better or worse than the case we have discussed?

Useful Links

1. Fundamentals of a Uniform Linear Array (ULA)
https://www.raymaps.com/index.php/fundamentals-of-a-uniform-linear-array-ula/
2. Basics of Beamforming in Wireless Communications
https://www.raymaps.com/index.php/basics-of-beam-forming-in-wireless-communications/
3. Multicarrier Beamforming at mmWave
https://www.raymaps.com/index.php/multicarrier-beamforming-at-mmwave/
4. Rectangular Array—Mathematical Model and Code
https://www.raymaps.com/index.php/fundamentals-of-a-rectangular-array-mathematical-model-and-code/
5. Circular Array—Mathematical Model and Code
https://www.raymaps.com/index.php/fundamentals-of-a-circular-array-mathematical-model-and-code/
6. Direction of Arrival Estimation
https://www.raymaps.com/index.php/direction-of-arrival-estimation/
7. Fundamentals of Linear Array Processing
https://www.raymaps.com/index.php/fundamentals-of-linear-array-processing-receive-beamforming/

References

1. https://electronics.stackexchange.com/questions/288530/uniform-linear-array-ula-beamwidth-and-angular resolution-using-fft
2. Gozasht, F., Dadashzadeh, G.R., Nikmehr, S.: A comprehensive performance study of circular and hexagonal array geometries in the lms algorithm for smart antenna applications. Prog Electromagnet Res PIER **68**, 281–296 (2007)
3. Van Trees, H.L.: Optimum array processing: Part IV of detection, estimation and modulation theory. Wiley (2002)
4. https://www.comm.utoronto.ca/~rsadve/Notes/DOA.pdf
5. https://arxiv.org/pdf/2002.01588.pdf
6. https://escholarship.org/content/qt1w13p71p/qt1w13p71p_noSplash_9f69e592ae9b76ba11f4349daa613333.pdf
7. Kay, S.M.: Fundamentals of statistical signal processing—estimation theory. Prentice Hall (1997)
8. Hicks, J.: Novel approaches to overloaded array processing. Doctoral dissertation, Virginia Tech (2003)

Useful Links

1. Fundamentals of a Uniform Linear Array (ULA)
   [illegible]
2. [illegible]
   [illegible]

Chapter 7
Phase and Frequency

7.1 Introduction

Finding the phase and frequency of a signal embedded in noise is a classical problem in signal processing. The problem becomes more complex when there are multiple signals spaced closely in the frequency domain. Some of the methods that we discuss in this chapter are Zero Crossing Detector, Kay's estimator, FFT, MUSIC and ESPRIT. The first two methods are quite accurate and easy to implement, but they work only for a single sinusoid embedded in noise. For more complex scenarios we have to use FFT, MUSIC or ESPRIT. The last method is particularly useful when there are multiple closely spaced sinusoids embedded in noise. But this method breaks down in the presence of high noise. The performance of MUSIC and ESPRIT also depends upon the model order used. Cramer Rao Lower Bound (CRLB) is an important benchmark against which you can compare the performance of your algorithm. If you have attained CRLB there is no need to go any further, this is the best you can do.

In modern times the most powerful technique that has been used by scientists and engineers to find the spectral content of a time series is Fast Fourier Transform, attributed to Cooley and Tukey, who jointly published a paper on this in 1965. But now it is widely accepted that the algorithm was first proposed around 1805 by Carl Friedrich Gauss, who used it to interpolate the trajectories of the asteroids Pallas and Juno. But his work was not widely recognized back then. Because of the importance of the algorithm several variations have been proposed over the years which target different fields and applications. The most attractive feature of FFT is that it makes the DFT and IDFT to be feasible by reducing the number of operation from $O\left(N^2\right)$ to $O(N\log N)$. Following are the relationships that describe the time and frequency resolution of FFT in terms of the sampling frequency f_s and temporal window T.

Y. Ahmed, *Recipes for Communication and Signal Processing*,
Signals and Communication Technology,
https://doi.org/10.1007/978-981-99-2917-7_7

$$\Delta t = 1/f_s$$

$$\Delta f = 1/T$$

We also discuss the Phase Lock Loop (PLL) which is a very important component of modern communication receivers where it is used for phase and frequency synchronization. It is also used in test and measurement equipment for frequency synthesis. It has three important components including a mixer, a low-pass filter and a Voltage Controlled Oscillator. The frequency of the output signal can be varied by controlling the frequency of the input or by the frequency of the feedback signal resulting in multiples of the base frequency (both lower and higher frequencies can be generated). We discuss two types of PLLs, namely Type-1 and Type-2 which are named so after the number of integrators they have. The performance of the Type-2 PLL is much better than Type-1 in terms of settling time and capture range.

PLLs are also sometimes used as FM demodulators when the input to the VCO is taken as the output. Remember that input to the VCO is a filtered error signal. The filter removes the sum frequency and allows the difference frequency to pass through resulting in a demodulated signal at the output. Lastly, remember that phase and frequency are inter-related as instantaneous frequency is just the rate of change of phase or we can say phase is the integral of the instantaneous frequency. Even a small error in instantaneous frequency such as 40 ppm for a 1 GHz signal (or 40 kHz) can cause a very high bit error rate if the resynchronization is not performed. But sometimes the error is cyclical, drifting to the positive side and then to the negative, resulting in a very small cumulative effect.

7.2 Modeling Phase and Frequency Synchronization Error

Carrier phase or frequency synchronization is a common problem in wireless communication systems. These two problems are inter-related as instantaneous frequency is just the rate of change of phase. The problem of carrier frequency offset might appear due to one of two reasons. Either the oscillators at the transmitter and receiver are not aligned in the frequency domain or there is a Doppler shift introduced by the channel (remember that a moving object in the wireless environment introduces a Doppler shift). In the case of the former the frequency misalignment is given in parts per million (ppm). A typical value for commercially available oscillators is $\pm$ 20 ppm. Assuming that there is maximum frequency error at both the transmitter and receiver, the error increases to $\pm$ 40 ppm. At 1 GHz this translates to 40 * 1,000,000,000/1,000,000 = 40 kHz.

MATLAB code for calculating bit error rate (BER) in presence of phase and frequency error is given below.

```
%%%%%%%%%%%%%%%%%%%%%%%%%%%%%%%%%%%%%%%%
%         BPSK MODULATION WITH CARRIER
%              PHASE/FREQUENCY ERROR
%           I is the number for BPSK symbols
%         N is the number of samples per symbol
%            fc is the carrier frequency in Hz
%           fs is the sampling frequency in Hz
%              Copyright 2020 RAYmaps
%%%%%%%%%%%%%%%%%%%%%%%%%%%%%%%%%%%%%%%%
clear all
close all

I=5e3;
N=40;
fc=1e9;
fs=10*fc;
ts=1/fs;
t=0:ts:ts*(N*I-1);
Eb=N;
EbNodB=10;
dp=0;
df=10e3;

% CARRIER GENERATION
tx_carrier=(sqrt(2))*cos(2*pi*fc*t);
rx_carrier=(sqrt(2))*cos(2*pi*(fc+df)*t+dp);

% SIGNAL GENERATION
tx_symbol=2*(round(rand(1,I))-0.5);
oversampled_sym_matrix=ones(N,1)*tx_symbol;
oversampled_sym_vector=reshape(oversampled_sym_matrix,1,N*I);
tx_signal=oversampled_sym_vector.*tx_carrier;

% NOISE ADDITION
EbNo=10^(EbNodB/10);
calibrated_noise=sqrt(Eb/(2*EbNo))*randn(1,N*I);
rx_signal=tx_signal+calibrated_noise;

% SIGNAL RECOVERY
down_converted_signal=rx_signal.*rx_carrier;
filtered_signal=ones(1,N)*reshape(down_converted_signal,N,I);
rx_symbol=sign(filtered_signal);

% BIT ERROR RATE CALCULATION
ber=(I-sum(tx_symbol==rx_symbol))/I
%%%%%%%%%%%%%%%%%%%%%%%%%%%%%%%%%%%%%%%%
```

Running the above code for 5000 BPSK symbols with a frequency error of 10 kHz at a carrier frequency of 1 GHz we get 7.5 errors per 1000 symbols passed through the channel. This is for at $E_b/N_o = 10\text{dB}$, where without any frequency or phase error, the bit error rate (BER) of BPSK should be 4 bits in a million bits passed through the channel, i.e., with frequency error there is more than 1000 times increase in BER. Lastly, we note that the bit errors increase as we increase the length of packet sent through the channel, as phase error cumulates. This can be thought of as the length of time before which the frequency synchronization has to be performed.

Note:

1. We have used passband simulation unlike our previous posts since the above concepts are easier to explain this way. But the downside is that simulation takes longer, since we sample at ten times the carrier frequency to get accurate results.
2. Please note that a constant frequency error increases the phase error and the results are quite adverse even for a small frequency shift. The cumulative phase

shift depends upon how long a symbol sequence we are simulating (since phase is just the integral of the frequency).
3. A drift in the local oscillator can be on either side, higher or lower, or even cyclic. So, it is possible that cumulative effect is very small.
4. In practical systems phase of the carrier is estimated at the receiver by a Phase Locked Loop (PLL). This is something we will discuss in the future post.

7.3 Frequency Estimation Using Zero Crossing Method

A sinusoidal signal is the most fundamental type of signal that exists in communication systems, power systems, navigation systems, etc. It is controlled by three parameters which are the amplitude, phase and frequency. The last two, that is phase and frequency, are interconnected. As discussed in my previous post Instantaneous Frequency (IF) is nothing but the rate of change of phase. This can be mathematically described as

$$\mathrm{IF} = \Delta\phi/\Delta t$$

It is sometimes important in communication systems to estimate the phase and frequency of the signal arriving at the receiver. This is especially true for Phase Modulated (PM) and Frequency Modulated (FM) systems. But any synchronous communication system requires that the carrier phase and frequency is synchronized between the transmitter and receiver. For more on carrier phase and frequency synchronization error please visit the previous section.

There are number of ways to estimate the frequency of a sine wave, from Fast Fourier Transform (FFT), to Autoregressive (AR) methods, to high resolution spectral estimation methods such as MUSIC* and ESPRIT**. The challenge here is to perform an accurate estimation with minimum number of samples and in presence of high noise and interference. Sometimes it is also required to detect multiple sinusoids simultaneously, which are overlapping in time domain and closely spaced in the frequency domain.

In this post we present the simplest method of frequency estimation that is called the Zero Crossing (ZC) method. Since a sine wave crosses the x-axis twice during each cycle, we can simply count the number of crossings and divide it by two and again divide it by the observation window size, giving us the frequency in Hertz. Please note that for this scheme to work we need to have a few complete cycles. Higher the length of the observation window greater is the accuracy.

It has been shown in [1] that at high signal to noise ratio (SNR > 10 dB) ZC estimator approaches the Cramer Rao Lower Bound (CRLB). MATLAB code for the ZC method is given below. For comparison we have also included the FFT method in our simulation, and results are shown in the figure below. Please remember that accuracy of the FFT method depends upon the window size T in the time domain. Mathematically, the frequency bin size is given as

$$\Delta f = \frac{1}{T}$$

To separate two sinusoids the minimum separation must be $2\Delta f$ and not Δf as there needs to be a vacant bin between the two sinusoids. Also, we experimented with the frequency resolution of Zero Crossing detector, and it was found out to be Δf /2 (although, unlike FFT, this method can detect only one tone at a time). It must be noted that frequency resolution of FFT does not increase by increasing the sampling frequency. In fact, if sampling frequency is increased and number of samples remains the same, frequency resolution decreases (Fig. 7.1).

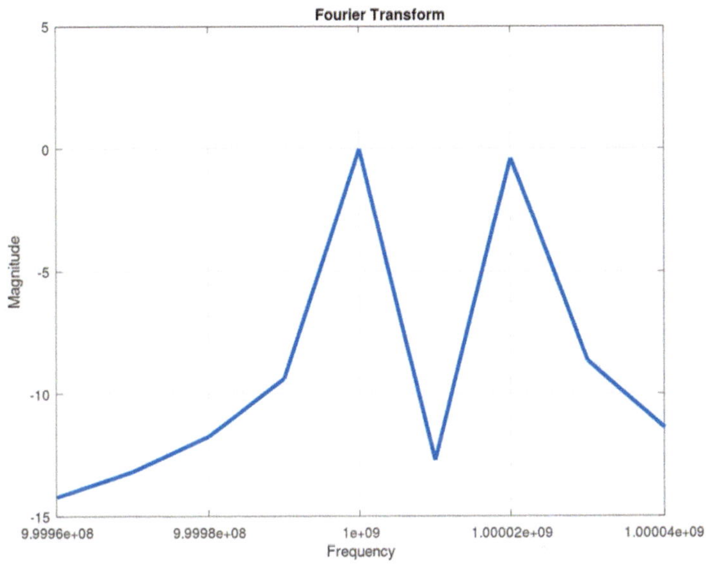

Fig. 7.1 FFT of two tones embedded in noise. *Note* The frequency estimate of zero crossing method can be improved greatly by using a simple moving average (MA) filter before the detection of zero crossings

```
%%%%%%%%%%%%%%%%%%%%%%%%%%%%%%%%%%%%%%%
%                    FREQUENCY ESTIMATION
%                           USING
%                    ZERO CROSSING METHOD
%                   N is the number of samples
%                 fc is the carrier frequency in Hz
%                fs is the sampling frequency in Hz
%                 dp is the phase error in radians
%                  df is the frequency error in Hz
%                    Copyright 2020 RAYmaps
%%%%%%%%%%%%%%%%%%%%%%%%%%%%%%%%%%%%%%%
clear all
close all

N=1e6;
fc=1e9;
fs=10*fc;
ts=1/fs;
t=0:ts:N*ts;
dp=pi/6;
df=20e3;

tx_carrier=sqrt(2)*cos(2*pi*fc*t);
rx_carrier=sqrt(2)*cos(2*pi*(fc+df)*t+dp)+0.1*randn(1,N+1);

% Frequency Estimation at Transmitter
count=0;
for n=1:N
  if tx_carrier(n)>0 && tx_carrier(n+1)<0  % -ve going
    count=count+1;
  end
  if tx_carrier(n)<0 && tx_carrier(n+1)>0  % +ve going
    count=count+1;
  end
end
tx_freq_estimate=count/2/t(end)

% Frequency Estimation at Receiver
count=0;
for n=1:N
  if rx_carrier(n)>0 && rx_carrier(n+1)<0  % -ve going
    count=count+1;
  end
  if rx_carrier(n)<0 && rx_carrier(n+1)>0  % +ve going
    count=count+1;
  end
end
rx_freq_estimate=count/2/t(end)
% Fourier Transform
two_tones=tx_carrier+rx_carrier;
fft_two_tones=abs(fft(two_tones));
fft_two_tones=fft_two_tones/max(fft_two_tones);
f=0:1/t(end):(N)*(1/t(end));
plot(f,10*log10(fft_two_tones),'linewidth',3);
axis([fc-2*df fc+2*df -15 5])
xlabel('Frequency')
ylabel('Magnitude')
title('Fourier Transform')
grid on
%%%%%%%%%%%%%%%%%%%%%%%%%%%%%%%%%%%%%%%
```

Complex Exponential

Sometimes the signal is not real (sinusoid) but is complex in nature (exponential). For such a case the above code can be slightly modified to do the Zero Crossing calculation. The method we adopt is to do the Zero Crossing calculation for both the real and imaginary parts separately and then take the average. The noise that is added to the signal is also complex in nature now. Last thing we want to mention is that applying a moving average filter just before the ZC algorithm greatly improves the estimate. The MATLAB/Octave code for this is given below.

```
%%%%%%%%%%%%%%%%%%%%%%%%%%%%%%%%%%%%%
%           FREQUENCY ESTIMATION USING
%         ZERO CROSSING METHOD (COMPLEX)
%             N is the number of samples
%            fc is the carrier frequency in Hz
%           fs is the sampling frequency in Hz
%            dp is the phase error in radians
%            df is the frequency error in Hz
%              Copyright 2020 RAYmaps
%%%%%%%%%%%%%%%%%%%%%%%%%%%%%%%%%%%%%
clear all
close all

N=1000;
fc=1e9;
fs=10*fc;
ts=1/fs;
t=0:ts:N*ts;
dp=2*pi*(rand-0.5);
df=1e6*(rand-0.5);

white_noise=sqrt(0.1/2)*(randn(1,N+1)+i*randn(1,N+1));
tx_carrier=exp(i*2*pi*fc*t);
rx_carrier=exp(i*2*pi*(fc+df)*t+dp)+white_noise;
rx_carrier=conv(rx_carrier, ones(1,5),'same');
```

```
% Frequency Estimation at Receiver (Real Part)
count1=0;
for n=1:N
  if real(rx_carrier(n))>0 && real(rx_carrier(n+1))<0  % -ve going
    count1=count1+1;
  end
  if real(rx_carrier(n))<0 && real(rx_carrier(n+1))>0  % +ve going
    count1=count1+1;
  end
end

% Frequency Estimation at Receiver (Imaginary Part)
count2=0;
for n=1:N
  if imag(rx_carrier(n))>0 && imag(rx_carrier(n+1))<0  % -ve going
    count2=count2+1;
  end
if imag(rx_carrier(n))<0 && imag(rx_carrier(n+1))>0  % +ve going
    count2=count2+1;
  end
end

rx_freq_estimate=mean([count1/2/t(end) count2/2/t(end)]);
freq_error=(fc+df)-rx_freq_estimate
%%%%%%%%%%%%%%%%%%%%%%%%%%%%%%%%%%%%%
```

* MUSIC: Multiple Signal Classification

** ESPRIT: Estimation of Signal Parameters via Rotational Invariance Technique

7.4 A Comparison of FFT, MUSIC and ESPRIT Methods of Frequency Estimation

As discussed in previous posts it is frequently required in communications and signal processing to estimate the frequency of a signal embedded in noise and interference. The problem becomes more complicated when the number of observations (samples) is quite limited. Typically, the resolution in the frequency domain is inversely proportional to the window size in the time domain. Sometimes the signal is composed of multiple sinusoids where the frequency of each needs to be estimated separately. Simple techniques such as Zero Crossing Estimator fail in such a scenario. Even some advanced techniques such as MATLAB function "pwelch" fail to distinguish closely spaced sinusoids.

In this post we discuss two of the most popular subspace methods of frequency estimation, namely MUSIC* and ESPRIT**. As a reference FFT method is also presented. Both MUSIC and ESPRIT work by separating the noisy signal into signal subspace and noise subspace [2, 4]. MUSIC works by exploiting the fact that noise eigen vectors that compose the noise subspace are orthogonal to the signal vectors (or steering vectors). So, we can search for the signal vectors that are most orthogonal to the noise eigen vectors and that gives us a frequency estimate. MUSIC is in fact an advanced form of Pisarenko Harmonic Decomposition (PHD) where the model order is one more than the number of sinusoids [3]. There is no such limitation in MUSIC.

Unlike MUSIC which exploits the noise subspace, ESPRIT method uses the signal subspace. It estimates the signal subspace S from the estimate of the signal correlation matrix R (same correlation matrix that was used to estimate noise subspace in MUSIC algorithm). Here is the trick you need to perform on the eigen vectors forming the signal subspace S. Split the matrix S into two staggered matrices $S1$ and $S2$ of size $(M - 1) \times p$ each, where M is the model order and p is the number of sinusoids. $S1$ is matrix S without the last row, and $S2$ is the matrix S without the first row. Now divide the second matrix $S2$ by $S1$ using the Least Squares (LS) approach to obtain matrix P. The angles of eigen values of P give us the estimate of the angular frequency (Figs. 7.2 and 7.3).

It is seen that frequency resolution of FFT and MUSIC methods depends upon the size of the temporal window. Greater the length of the time window higher is the frequency resolution. In the results we have shown the frequency resolution is 4 MHz. But there is no such barrier for the ESPRIT method. In fact, in the absence of noise, we have observed that ESPRIT method can even decipher sinusoids placed 0.1 MHz apart. But once noise is added the frequency estimation of ESPRIT is not that great. So, for this method the limiting factor is the AWGN noise added to the signal not the size of the temporal window. Lastly, for the simulation scenario we have considered the model size M for MUSIC and ESPRIT is set to 100. We would further investigate this in the future post.

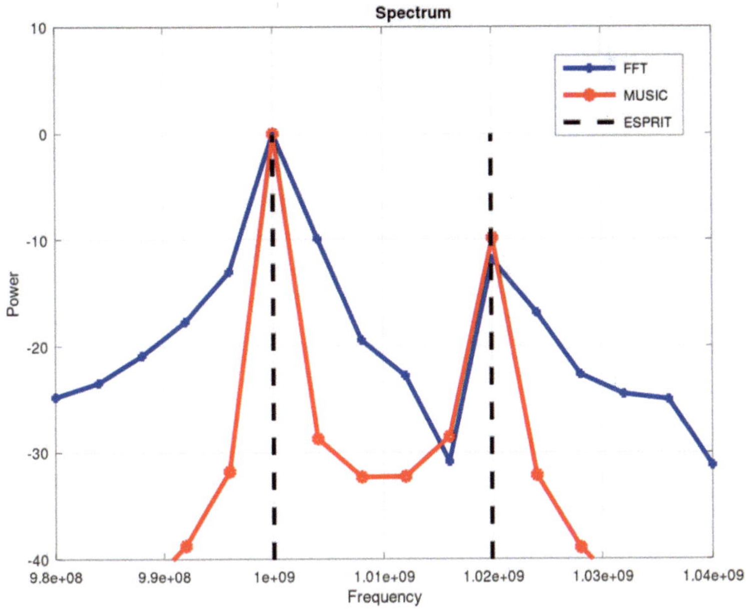

Fig. 7.2 PSD of two tones 20 MHz apart, SNR = 7 dB, SIR = 10 dB

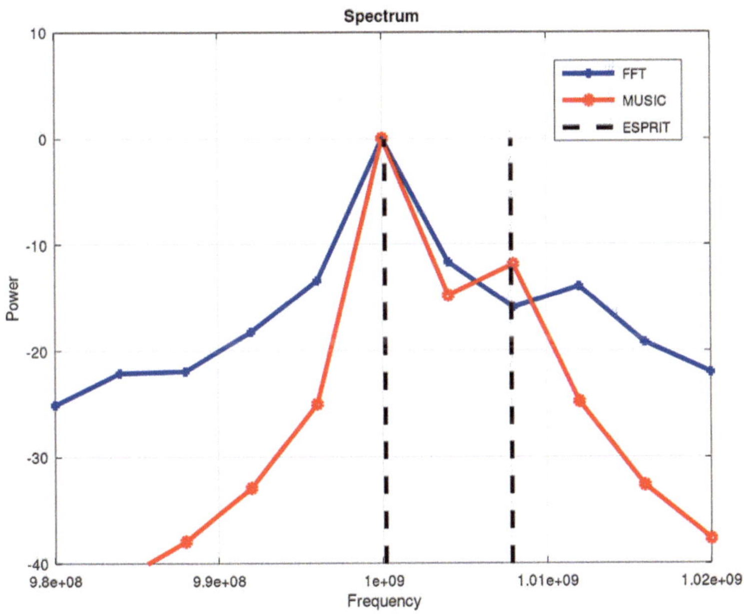

Fig. 7.3 PSD of two tones 8 MHz apart, SNR = 7 dB, SIR = 10 dB

```
%%%%%%%%%%%%%%%%%%%%%%%%%%%%%%%%%%%%%%%%%%%%%%%%%%
%                    SPECTRAL ESTIMATION USING
%                    FFT, MUSIC/PISARENKO, ESPRIT
%                    N is the number of samples
%                    f1 is the frequency of the signal in Hz
%                    f2 is the frequency of the interferer in Hz
%                    fs is the sampling frequency in Hz
%                    Copyright 2020 RAYmaps
%%%%%%%%%%%%%%%%%%%%%%%%%%%%%%%%%%%%%%%%%%%%%%%%%%
clear all
close all

N=1001;
f1=1.0000e9;
f2=1.0200e9;
fs=4*f1;
w1=2*pi*f1/fs;
w2=2*pi*f2/fs;

%%%%% Signal Generation in AWGN Noise %%%%%
n=0:N-1;
s1=sqrt(1.00)*exp(i*w1*n);
s2=sqrt(0.10)*exp(i*w2*n);
wn=sqrt(0.10)*(randn(1,N)+i*randn(1,N));
x=s1+s2+wn;
x=x(:);
```

```
%%%%% FFT Spectrum Generation %%%%%
f=0:fs/(N-1):fs;
FFT_abs=abs(fft(x));
plot(f,20*log10(FFT_abs/max(FFT_abs)),'linewidth',3,'b+-');
hold on

%%%%% MUSIC/Pisarenko Spectrum Generation %%%%%
M=100;
p=2;

f=0:fs/(N-1):fs;
w=2*pi*f/fs;

cov_matrix=zeros(M,M);
for n=1:N-M+1
  cov_matrix=cov_matrix+(x(n:n+M-1))*(x(n:n+M-1))';
end
[V,lambda]=eig(cov_matrix);
e=exp(j*((0:M-1)')*w);

den=zeros(length(w),1);
for n=1:M-p
  v=(V(:,n));
  den=den+(abs((e')*v)).^2;
end
PSD_MUSIC=1./den;
plot(f,10*log10(PSD_MUSIC/max(PSD_MUSIC)),'linewidth',3,'ro-')
hold on

%%%%% ESPRIT Based Frequency Calculation %%%%%
M=100;
p=2;

cov_matrix=zeros(M,M);
for n=1:N-M+1
  cov_matrix=cov_matrix+(x(n:n+M-1))*(x(n:n+M-1))';
end

[U,E,V]=svd(cov_matrix);
S=U(:,1:p);
S1=S(1:M-1,:);
S2=S(2:M,:);
P=S1\S2;
w=angle(eig(P));
f_est=fs*w/(2*pi);

%%%%% Plotting the PSD %%%%%
line([f_est(1), f_est(1)],[-40, 0],'Color','Black','LineStyle','--','linewidth',3);
line([f_est(2), f_est(2)],[-40, 0],'Color','Black','LineStyle','--','linewidth',3);
title('Spectrum')
xlabel('Frequency')
ylabel('Power')
legend('FFT','MUSIC','ESPRIT')
axis([0.98e9 1.04e9 -40 10])
grid on
hold off
%%%%%%%%%%%%%%%%%%%%%%%%%%%%%%%%%%%%%%%%%%%%%%%%
```

Note:

1. ESPRIT method can be thought of as differential demodulation of a Continuous Phase Modulated (CPM) signal. But instead of working with staggered samples we are working with staggered matrices. It's the rotation that we want to measure in both.
2. One point to be noted is that both MUSIC and ESPRIT methods require the knowledge of number of sinusoids embedded in noise. FFT on the other hand does not need this information beforehand.

3. From the code it may seem that noise variance is 0.1 but it's actually 0.1 per dimension. The total noise variance is 0.2 resulting in an SNR of 10 * log10(1/0.2) = 7 dB.
4. At a sampling frequency of 4 GHz, 1001 samples translate to a time window of only 250nsec. Thus, the resolution of FFT method is 1/250nsec = 4 MHz.
5. Please note that the Zero Crossing algorithm we presented in the last post can be easily modified to work with complex exponentials. After all, a complex exponential can be broken down into real and imaginary parts and Zero Crossings of each can be calculated separately.

*MUSIC: Multiple Signal Classification
**ESPRIT: Estimation of Signal Parameters via Rotational Invariant Techniques

7.5 KAY's Single Frequency Estimator

As previously discussed, finding the frequency of a complex sinusoid embedded in noise is a classical problem in signal processing. The problem is compounded by the fact that number of samples available is usually quite small. So far, we have discussed Zero Crossing, FFT, MUSIC and ESPRIT methods of frequency estimation. Zero Crossing method is simplest of the above four, but it can detect only one sinusoid at a time. Advantage of Zero Crossing method is that it is computationally not that complex. It does not require complex matrix manipulations as some of the other methods do.

Now we present another single frequency estimator that is quite easy to implement. It is known as Kay's estimator [5] after its founder Steven M. Kay. Although quite simple to implement it achieves Cramer Rao Lower Bound (CRLB) for moderate to high SNRs (SNR > 10 dB). The basic concept of Kay's estimator is quite simple. It finds the phase advanced from one sample to the next, and this gives us the frequency. Averaging is performed over the time window to get more accurate results. Averaging can be performed using uniform weighting or a weighting function defined by Kay in [5]. In the latter case CRLB is achieved with equality, so there is no need to look for another estimator.

Mathematically CRLB is given as
Mean Squared Error (MSE) in Angular Frequency = 6/(SNR * N * (N^2–1)).

where N is the number of samples (set to 100 in this case) and SNR is the signal to noise ratio (set to 10 dB). MATLAB/Octave code for both the methods (with different weighting functions) is given below. For the example given CRLB is calculated as.

$$\mathrm{MSE} = 6/(10 * 100 * (100^2 - 1)) = -62.22\,\mathrm{dB}$$

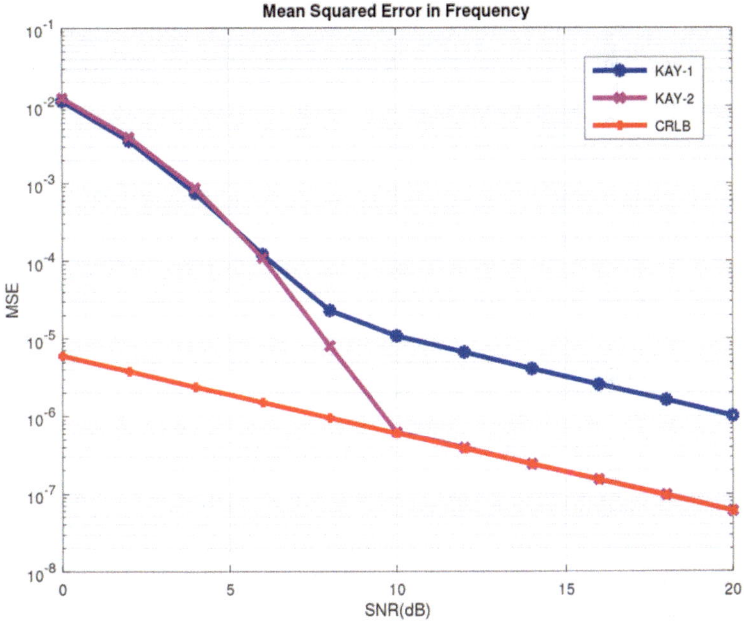

Fig. 7.4 MSE of KAY frequency estimation methods as a function of SNR

```
%%%%%%%%%%%%%%%%%%%%%%%%%%%%%%%%%%%%
%          FREQUENCY ESTIMATION USING
%                   KAY's ESTIMATOR
%             (SINGLE FREQUENCY ONLY)
%             N is the number of samples
%           fc is the carrier frequency in Hz
%          fs is the sampling frequency in Hz
%               Copyright 2020 RAYmaps
%%%%%%%%%%%%%%%%%%%%%%%%%%%%%%%%%%%%
clear all
close all

N=100;
fc=1.0e9;
fs=20*fc;
w=2*pi*fc/fs;

n=0:N-1;
s=sqrt(1.0)*exp(i*w*n);
wn=sqrt(0.1/2)*(randn(1,N)+i*randn(1,N));
x=s+wn;

% KAY ESTIMATOR-1 (uniform weighting)
w_est=0;
M=N;
for m=1:M-1
  w_est=w_est+(1/(M-1))*arg((x(m))'*x(m+1));
end
error_1=w-w_est;

% KAY ESTIMATOR-2 (using weighting function)
w_est=0;
M=N;
for m=1:M-1
  p(m)=6*(m)*(M-m)/(M*(M^2-1));
  w_est=w_est+p(m)*arg((x(m))'*x(m+1));
end
error_2=w-w_est;
%%%%%%%%%%%%%%%%%%%%%%%%%%%%%%%%%%%%
```

Note:

1. The complete code for comparing the MSE of the two methods with CRLB is left as a homework assignment.
2. Please note that as the number of samples N is increased the CRLB and Mean Squared Error (MSE) both go down. But after a time, the law of diminishing returns sets in and MSE does not go down as rapidly as the CRLB.
3. The MSE also depends upon number of samples per cycle and having a very small number does not help in reducing the MSE. A few complete cycles work the best (in our code we use five complete cycles to do the estimation) (Fig. 7.4).

7.6 Phase Lock Loop—Explained

Phase Lock Loops (PLLs) are an important component of communication systems, where they are used for carrier phase and frequency synchronization. They are also used in test and measurement equipment such as in Signal Generators and Vector Network Analyzers (VNAs) for frequency synthesis [6–8]. Although not discussed here in detail but PLLs are also quite adept at generating multiples of a base frequency, e.g., if you have a reference signal at 10 MHz, then a PLL can be used to generate a 100 MHz signal ($X = 10$) or even a 1 GHz signal ($X = 100$). In fact, you can also divide the frequency to get low frequency signals. In the first case the feedback frequency is divided by X, and in the second case the reference or input frequency is divided by X.

All Analog Phase Lock Loops have three basic components: a mixer, a low-pass filter (LPF) and a Voltage Controlled Oscillator (VCO) [1]. The inputs to the mixer include a reference signal (i/p signal) and a VCO signal (o/p signal). In case these are not phase and frequency synchronized an error signal will be generated which will consist of a sinusoid that has a frequency which is the difference of the two frequencies and a sinusoid that has a frequency which is the sum of the two frequencies. An LPF is then used to remove the high frequency component, and the low frequency component is used as a control signal by the VCO. As the name suggests the output of the VCO is a signal whose frequency is directly proportional to the input voltage. Higher the difference between input and output frequency greater is the value of the error signal fed to the VCO.

There are three main modes of operation of PLLs [6] (Figs. 7.5, 7.6, 7.7 and 7.8).

1. Free running: No input is applied
2. Capture mode: Input is applied and output starts tracking it
3. Lock mode: Input and output are synchronized in phase and frequency

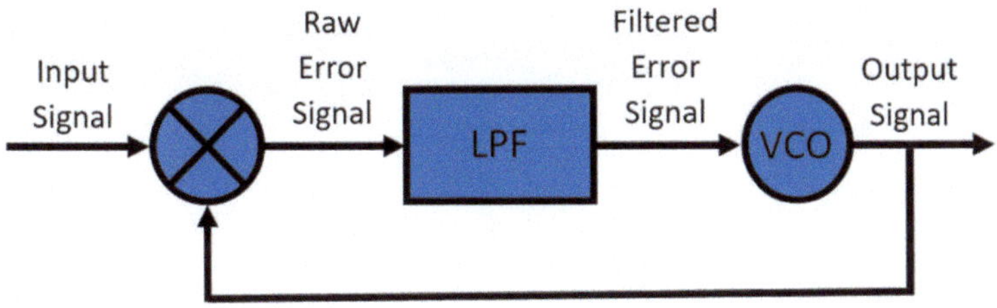

Fig. 7.5 Block diagram of Type-1 Phase Lock Loop

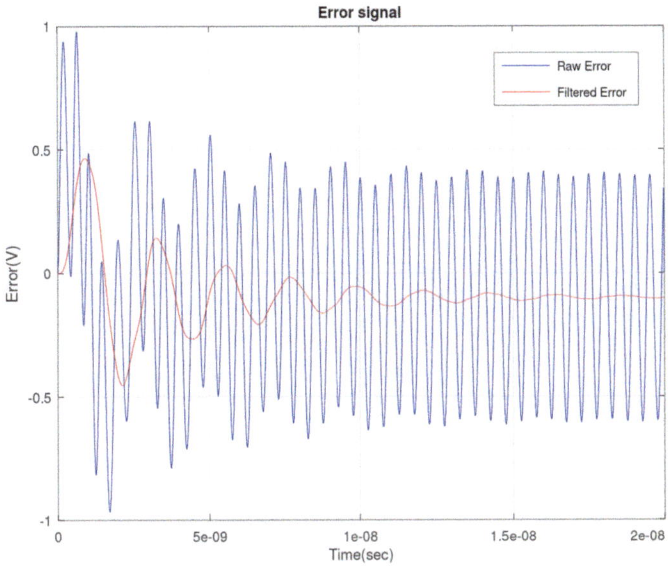

Fig. 7.6 Error signal of Type-1 PLL

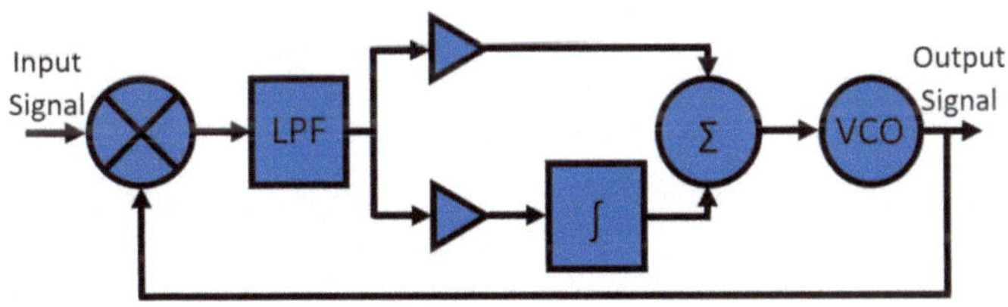

Fig. 7.7 Block diagram of Type-2 Phase Lock Loop

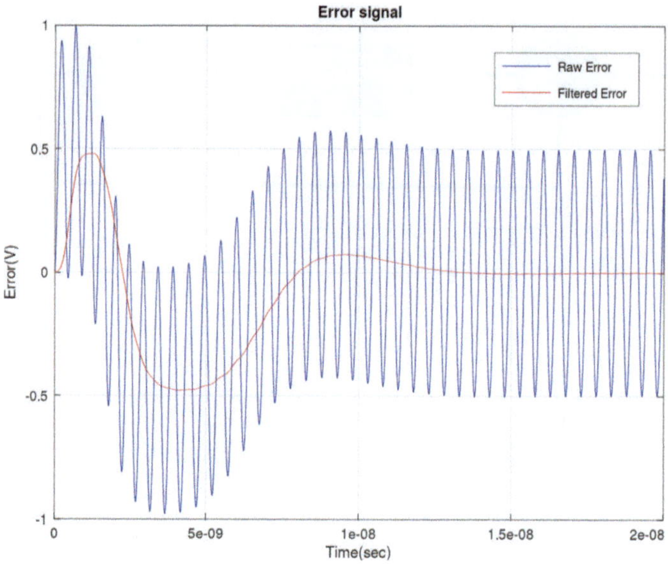

Fig. 7.8 Error signal of Type-2 PLL

Type-1 Phase Lock Loop

```
%%%%%%%%%%%%%%%%%%%%%%%%%%%%%%%%%%%%%%%%%%%%%%%%%%%%%%%%%%%%%%%%%%%%%%%%
%            Sampled time-domain simulation of an analog Phase Locked Loop (PLL)
%                 (Type-1 with Multiplier, Low-Pass Filter and VCO)
%                          Adapted from aaronscher.com
%
%                 The input signal or reference signal is a simple sinusoid.
%               The output signal is also a sinusoid that tracks the frequency
%                  of the reference signal after a certain start up time.
%                              N is the number of samples
%                            fc is the carrier frequency in Hz
%                           fs is the sampling frequency in Hz
%                       f_VCO is the VCO reference frequency in Hz
%                                K_VCO is the VCO gain
%                              Copyright 2020 RAYmaps
%%%%%%%%%%%%%%%%%%%%%%%%%%%%%%%%%%%%%%%%%%%%%%%%%%%%%%%%%%%%%%%%%%%%%%%%
clear all
close all

%Simulation Parameters
N=2000;
fc=1e9;
phase_ref=0.5;
f_VCO=1.1e9;
K_VCO=1.0e9;
fs=100*fc;
filter_length=100;
```

```
% Defining the Filter Coefficients
filter_coeff=0.54-0.46*cos(0:2*pi/filter_length:2*pi);
filter_coeff=filter_coeff/sum(filter_coeff);

% Initialiazing the Signal Vectors
ts=1/fs;
t_vec=0:ts:(N-1)*ts;
ref_signal=sin(2*pi*fc*t_vec+phase_ref);
output_phase=zeros(1,N);
output_signal=zeros(1,N);
error_mult=zeros(1,N+filter_length);

% PLL Loop (Multiplier, Filter, VCO)
for n=2:N
  t=(n-2)*ts;
  error_mult(n+filter_length-1)=ref_signal(n)*output_signal(n-1);
  error_filtered(n)=sum(error_mult(n-1:n+filter_length-1).*(filter_coeff));
  output_phase(n)=output_phase(n-1)+2*pi*error_filtered(n)*K_VCO*ts;
  output_signal(n)=sin(2*pi*f_VCO*t+output_phase(n));
end

%Plot Error Signal
plot(t_vec,error_mult(filter_length+1:end),'b');hold on
plot(t_vec,error_filtered,'r');hold off
title('Error signal')
xlabel('Time(sec)')
ylabel('Error(V)')
legend('Raw Error','Filtered Error')
grid on
%%%%%%%%%%%%%%%%%%%%%%%%%%%%%%%%%%%%%%%%%%%%%%%%%%%%%%%%%%%%%%%%%%%%%%%%%%%%%%%%%%%%%%%%%%
```

Type-2 Phase Lock Loop

```
%%%%%%%%%%%%%%%%%%%%%%%%%%%%%%%%%%%%%%%%%%%%%%%%%%%%%%%%%%
%       Sampled time-domain simulation of an analog Phase Locked Loop (PLL)
%       Type-2 with Multiplier, Low-Pass Filter, VCO, Gains and Integrator
%       Adapted from aaronscher.com
%
%       The input signal or reference signal is a simple sinusoid.
%       The output signal is also a sinusoid that tracks the frequency
%       of the reference signal after a certain start up time.
%
%       N is the number of samples
%       fc is the carrier frequency in Hz
%       fs is the sampling frequency in Hz
%       ts is the sampling time in seconds
%       f_VCO is the VCO reference frequency in Hz
%       K_VCO is the VCO gain
%       G1 and G2 are the gains
%       Copyright 2020 RAYmaps
%%%%%%%%%%%%%%%%%%%%%%%%%%%%%%%%%%%%%%%%%%%%%%%%%%%%%%%%%%
clear all
close all

% Simulation Parameters
N=2000;
fc=1e9;
phase_ref=0.5;
f_VCO=1.1e9;
K_VCO=0.5e9;
fs=100*fc;
filter_length=100;
G1=0.5;
G2=2.0e8;

% Defining the Filter Coefficients
filter_coeff=0.54-0.46*cos(0:2*pi/filter_length:2*pi);
filter_coeff=filter_coeff/sum(filter_coeff);

% Initialiazing the Signal Vectors
ts=1/fs;
t_vec=0:ts:(N-1)*ts;
ref_signal=sin(2*pi*fc*t_vec+phase_ref);
output_phase=zeros(1,N);
output_signal=zeros(1,N);
branch1=zeros(1,N);
branch2=zeros(1,N);
error_mult=zeros(1,N+filter_length);

% PLL Loop (Multiplier, Filter, VCO)

for n=2:N
  t=(n-2)*ts;
  error_mult(n+filter_length-1)=ref_signal(n)*output_signal(n-1);
  error_filtered(n)=sum(error_mult(n-1:n+filter_length-1).*(filter_coeff));
  branch1(n)=G1*error_filtered(n);
  branch2(n)=branch2(n-1)+G2*error_filtered(n)*ts;
  PI_error(n)=branch1(n)+branch2(n);
  output_phase(n)=output_phase(n-1)+2*pi*PI_error(n)*K_VCO*ts;
  output_signal(n)=sin(2*pi*f_VCO*t+output_phase(n));
end

%Plot Error Signal
plot(t_vec,error_mult(filter_length+1:end),'b');hold on
plot(t_vec,error_filtered,'r');hold off
title('Error signal')
xlabel('Time(sec)')
ylabel('Error(V)')
legend('Raw Error','Filtered Error')
grid on
%%%%%%%%%%%%%%%%%%%%%%%%%%%%%%%%%%%%%%%%%%%%%%%%%%%%%%%%%%
```

Type-1 versus Type-2

We have seen that the filtered error of Type-2 PLL goes down much more rapidly than Type-1 PLL. But now let us look at the input to the VCO, the control signal to the VCO. This is not the same as LPF output as one more stage is added in Type-2 PLL.

But before we do an analysis of the results, we must warn you that a PLL is a nonlinear system, and the results might not always make sense. Some linear models of PLLs do exist that make the analysis simpler. This will be the subject of a future article.

If we make K_VCO to be the same for both the PLLs (it was not the same in the above examples) then we can make an apple-to-apple comparison. We set K_VCO to be 1.0e9 for both the cases and then look at the VCO control signal. As expected, the VCO control signal settles down much more quickly in Type-2 than in Type-1. This is the main advantage of Type-2 PLL; it locks on to the reference much more quickly (Fig. 7.9).

Simulation results have shown that capture range of Type-2 PLL is also larger than Type-1 PLL. When the difference in input and output frequency is more than 400 MHz the control voltage of Type-1 PLL keeps oscillating and PLL does not go into lock. One might think that reducing the cutoff frequency of the LPF can solve this problem, but this is not the case. According to some references increasing K_ VCO and cutoff frequency of the LPF in fact increases the capture range.

To further understand the PLL we need to revisit damped sinusoidal function and its s-domain equivalent. This will not be discussed in this post.

Note:

1. A PLL can also be used as an FM demodulator by taking the output of the loop filter before it's fed to the VCO.
2. It is possible for a PLL to have a phase offset between input and output, but when locked, the output frequency must exactly track the input frequency.
3. The type of a PLL refers to the number of poles of the loop transfer function located at the origin. This is controlled by the number of integrators in the loop, VCO being one integrator itself.
4. For the filter coefficients we just used a Hamming window since it works quite well and not everyone has access to signal processing toolbox which provides the "fir1" function.
5. For Digital PLLs the mixer is replaced by a XOR gate, which gives a high output whenever the two digital inputs are different. This signal is then passed on to a simple moving average filter the output of which is used to control the VCO. When the two inputs to the XOR gate are perfectly synchronized in phase and frequency the output of the XOR gate is zero.

Questions and Numerical Problems

1. What is the relationship between phase and frequency of a sinusoidal signal?
2. What is the resolution of Fast Fourier Transform in the frequency domain? Also, comment on the resolution of MUSIC and ESPRIT methods.

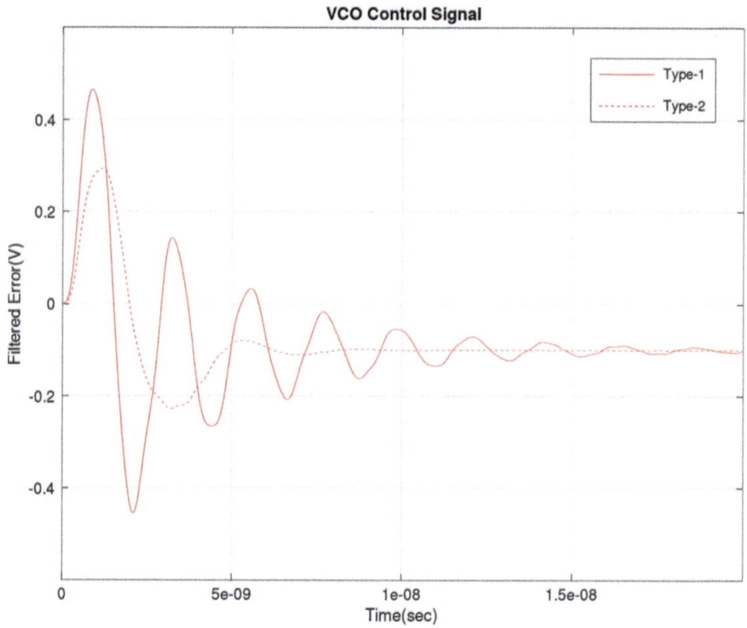

Fig. 7.9 VCO control signal Type-1 versus Type-2

3. Perform Discrete Fourier Transform (DFT) on a sinusoidal signal. Then perform Fast Fourier Transform (FFT) on the same signal. Do you notice any difference in computation time or accuracy (select a large enough temporal window)?
4. Do you see any similarity between DFT and Direction of Arrival (DOA) estimation using correlation method?
5. What are the main advantages and disadvantages of the Zero Crossing method of frequency estimation?
6. Simulate the Zero Crossing method of frequency estimation for a complex exponential embedded in noise. Compare the accuracy of frequency estimate with Cramer Rao Lower Bound (CRLB).
7. If the model order of Pisarenko Harmonic decomposition is three then how many sinusoids can be detected?
8. What property of signal subspace and noise subspace is used in the MUSIC algorithm? How does the MUSIC algorithm improve upon the Pisarenko Harmonic Decomposition?
9. What is the fundamental difference between MUSIC and ESPRIT methods of high resolution spectral estimation? Also, compare their accuracy and computational speed.
10. If the sampling frequency of a signal is increased ten times, but the total number of time domain samples remains the same, what impact will it have on the frequency resolution of the FFT method?
11. Compare the performance of Zero Crossing detector and Kay's estimator. Also, comment on the performance of both the methods with respect to CRLB.

12. Find the optimum number of time domain samples for Zero Crossing detector and Kay's estimator.
13. Let's assume a PLL has a reference signal of 10 MHz. How can it be used to generate a 1 MHz signal and a 100 MHz signal?
14. Explain in detail the role of each of the three components of the PLL, namely (a) mixer, (b) low-pass filter and (c) Voltage Controlled Oscillator.
15. Explain in detail the three modes of operation of the PLL, namely (a) free running, (b) capture mode and (c) lock mode.

Useful Links

1. Modeling Phase and Frequency Synchronization Error

   https://www.raymaps.com/index.php/modeling-phase-and-frequency-synchronization-error/
2. Frequency Estimation Using Zero Crossing Method

   https://www.raymaps.com/index.php/frequency-estimation-using-zero-crossing-method/
3. A Comparison of FFT, MUSIC and ESPRIT Methods of Frequency Estimation

   https://www.raymaps.com/index.php/a-comparison-of-fft-music-and-esprit-methods-of-frequency-estimation/
4. KAY's Single Frequency Estimator

   https://www.raymaps.com/index.php/kays-single-frequency-estimator/
5. Phase Lock Loop—Explained

   https://www.raymaps.com/index.php/phase-lock-loop-explained/

References

1. Liao, Y.: Phase and Frequency Estimation: High-Accuracy and Low-Complexity Techniques, MS Thesis Worcester Polytechnic Institute, May (2011)
2. https://en.wikipedia.org/wiki/MUSIC_(algorithm)
3. https://en.wikipedia.org/wiki/Pisarenko_harmonic_decomposition
4. https://en.wikipedia.org/wiki/Estimation_of_signal_parameters_via_rotational_invariance_techniques
5. Kay, S.M.: A Fast and accurate single frequency estimator. IEEE Trans. Acoust. Speech Signal Process. **37**(12) (1989)
6. http://www.aaronscher.com/phase_locked_loop/matlab_pll.html
7. https://www.electronics-notes.com/articles/radio/pll-phase-locked-loop/tutorial-primer-basics.php
8. https://www.analog.com/en/analog-dialogue/articles/phase-locked-loop-pll-fundamentals.html

Chapter 8
Advanced Topics

8.1 Introduction

As we come to a close of this book, we discuss some advanced topics, some new technologies which have already made it to the 5G standard and some which will probably see the implementation in 6G or later. We ask ourselves, what are the key resources for a modern wireless communication system that promises tens of gigabit per second of download speeds on the go or at home, in a multistory building in downtown New York, a suburban home or while traveling in a train at 600 km/h? As we have mentioned several times the key resources are the signal power and bandwidth. The more of these two resources you have, the higher is the data transfer speed. At 5G speeds a 4 k movie can be downloaded in less than a minute on the go. The latency of 5G is also substantially smaller than for any other standard, on the order of 1 ms or one thousandth of a second.

The power that can be transmitted and bandwidth consumed depends more upon regulatory approvals then on actual technology to be used. In the USA Federal Communication Commission (FCC) regulates communications by radio, television, wire, satellite and cable across the country. Telecom service providers in the USA have paid more than USD 100 billion in the 5G spectrum auction to the FCC. Since 5G would be using millimeter wave frequencies with cell phone towers which would be at street level, some concerns have been raised about exposure of human beings to the electromagnetic waves. The three most important parameters used to measure electromagnetic (EM) exposure are Specific Absorption Rate (SAR), power density and temperature increase. 5G experts argue that the 5G EM waves are non-ionizing and cause only localized heating of the tissue. No other adverse effects have been reported so far.

Along with power and bandwidth, the third dimension that can be used to increase the capacity of wireless communication systems is spatial dimension. It's not uncommon for modern wireless communication systems to use 64 antennas at the transmitter and 64 antennas at the receiver to make use of spatial domain. In a

Y. Ahmed, *Recipes for Communication and Signal Processing*,
Signals and Communication Technology,
https://doi.org/10.1007/978-981-99-2917-7_8

recent test by T-Mobile it has been shown that in an MU-MIMO configuration with 16 parallel streams a spectral efficiency of more than 50bps/Hz can be achieved. Massive MIMO and millimeter wave go hand in hand as the implementation of large arrays is only possible when wavelengths are the order of a few millimeters, e.g., an 8×8 array at a frequency of 30 GHz would be the size of 4 cm $\times$ 4 cm, something that would easily fit in the palm of your hand. In the future wireless communication systems we may see thousands of antennas deployed at a base station spread over a large geographical area such as on the facade of buildings.

Last three concepts we would like to introduce are Full Duplex, Reconfigurable Intelligent Surfaces and Index Modulation. Most wireless communication systems are either FDD or TDD, i.e., they transmit and receive at different frequencies or in different time slots. But what if we can transmit and receive over the same frequency and time slot, this can potentially double the data rate of a wireless system. The problem with this is that the transmitted signal, which is at much higher power level, feeds back into the receiver (transmit signal might be as much as 100 dB stronger). But this can be canceled using multiple antenna techniques or RF cancelation or digital cancelation. Reconfigurable Intelligent Surfaces are meta-surfaces that act like relays, that retransmit the signal but are passive, meaning they do not amplify the signal but neither do they add noise. Lastly, Index Modulation or Spatial Modulation is a technique where instead of transmitting from all the antennas of a MIMO transmitter, a select set is used. The index of the transmit antennas used also providing information to the receiver. It goes without saying that the spatial channels must be uncorrelated for this scheme to work.

8.2 Beyond Massive MIMO

Recently Björnson and Marzetta in their publication on antenna arrays [1] discussed five possible future research directions. In their opinion Massive MIMO is no longer a theoretical concept, and it is already being adopted in the industry. It is not uncommon to find 64 element antenna arrays being deployed in wireless communication systems. So, we now need to look beyond Massive MIMO or MaMIMO as it is popularly referred to. Here are three possible future research directions that we find most interesting.

Extremely Large Aperture Array

It is well known that spatial resolution of an array depends upon the aperture of the array or in simple terms the size of the array. So, in the future wireless communication systems we may see thousands of antennas deployed at a base station spread over a large geographical area (Fig. 8.1). As discussed in [1] one possibility is to deploy these antennas on the facade of buildings located in urban city centers. But a problem faced in such a configuration is that interelement spacing may be greater than half the wavelength causing spatial aliasing. This is similar to the concept of aliasing when an under-sampled signal is transformed from time domain to frequency domain. Also

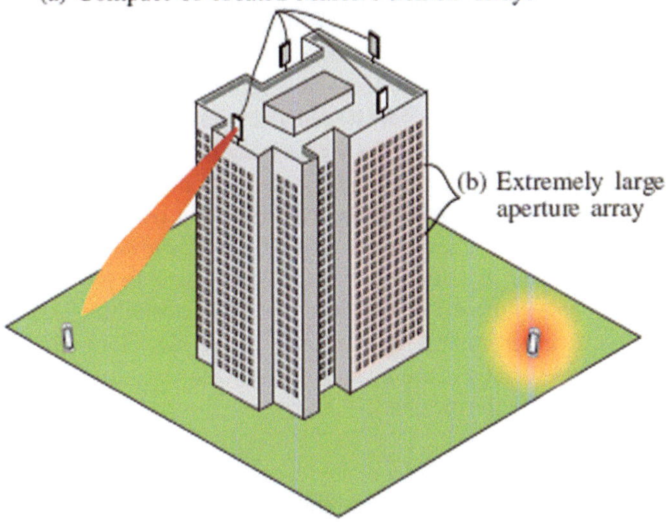

Fig. 8.1 Extremely large aperture arrays (Fig. 4 from [1])

note that the frequency resolution is dependent on the time domain window size and not sampling frequency as is popularly believed.

One solution to the problem of spatial aliasing is to have an aperture so large that the beam pattern is like a fine ray pointing in the direction of interest. Rays that may exist in other directions are so narrow that the chance of interfering with unwanted user is very low. Another problem is that far field assumption would not hold true as the mobile station distance from the base station (a building in this case) would be comparable to the size of the array. To be exact the far field distance of an antenna array is given as $d_f > 2D^2/\lambda$. So as the signal wavelength λ decreases (especially as we move to millimeter wave) and as the antenna dimension D increases the antenna far field distance increases. Lastly, we would like to point out that in all our discussions so far on antenna arrays we have considered a uniform linear array, but non-uniform geometries can provide better spatial resolution in some cases.

3D Spatial Location

Earlier wireless communication systems had very coarse spatial location (50–200 m) which was based on Cell ID and inaccurate timing measurements. But over the years the accuracy has been improved greatly with current systems achieving accuracy of about 10 m. This has been made possible by using GPS signals and more accurate Time Difference of Arrival (TDoA) techniques. Remember that GPS signals alone are not sufficient as they do not work indoors and in urban canyons. In the future wireless communication systems location accuracy would be improved to less than 10 m, and we would be able to define position along X, Y and Z. It is expected that it would also be possible to measure the roll, pitch and yaw of a mobile device (referred to as 6D Spatial Location). This would be made possible by large antenna arrays.

Large-Scale MIMO Radar

Radar or Radio Detection and Ranging was first discovered when an aircraft accidentally interfered in the signal transmission between a radio transmitter and a radio receiver. Eight countries started working on it at more or less the same time and had it ready before World War-II. MIMO radar is a relatively new concept, and the paper that most of the researchers like to cite is by E. Fishler from 2004 [2]. The concept of MIMO radar is to use multiple antennas at the transmitter and receiver to improve the detection of signals. These transmitters or receivers may or may not be collocated. The number of targets that a MIMO radar can detect is usually much higher than typical radar. A Large Aperture MIMO radar can provide much better spatial resolution and improved interference rejection capabilities. Another important concept that researchers have worked on is that of Waveform Diversity which allows multiple waveforms to be transmitted simultaneously from the transmitters and distinctly detected at the receivers.

Full Duplex: A Bonus Inclusion

Full Duplex (FD) was not a part of the original article but is now included due to its promise of doubling the capacity of wireless systems. It was originally thought that it would make it to 5G but the challenge of implementing it in real world scenarios still remains, and it will most likely get included in 6G or whatever the next generation of wireless systems is called. In simple terms most wireless communication systems work by transmitting at one frequency and receiving at another frequency, this is called FDD. Some wireless communication systems work by transmitting and receiving at the same frequency but having different time slots for each, this is called TDD. In Full Duplex (FD) the transmission and reception happens at the same frequency and within the same time slot [3]. The challenge here is that the transmitted signal feeds back into the receive chain, and the received signal is drowned in the transmitted signal.

Several techniques have been proposed in the literature to cancel this echo as it is sometimes called. They can be broadly classified into antenna-based techniques, RF techniques and digital techniques. Antenna techniques are the simplest as they propose to place the transmit antennas (usually 2) at an appropriate distance from the receive antenna so that the echoes cancel out on arrival (Fig. 8.2). Another technique simply inverts the phase of the transmitted signal and then adds it to the echo after adjusting the power level appropriately. Lastly are digital techniques that also apply some form of echo cancelation to recover the received signal. All these techniques are simple in theory but much harder to implement in practice where the wanted signal is drowned in 60–90 dB more powerful unwanted signal.

Note:

Although antenna cancelation is a very simple and attractive option it has certain weaknesses.

1. Full Duplex using antenna cancelation requires two transmit and one receive antenna. Full Duplex only doubles throughput, whereas a three antenna MIMO

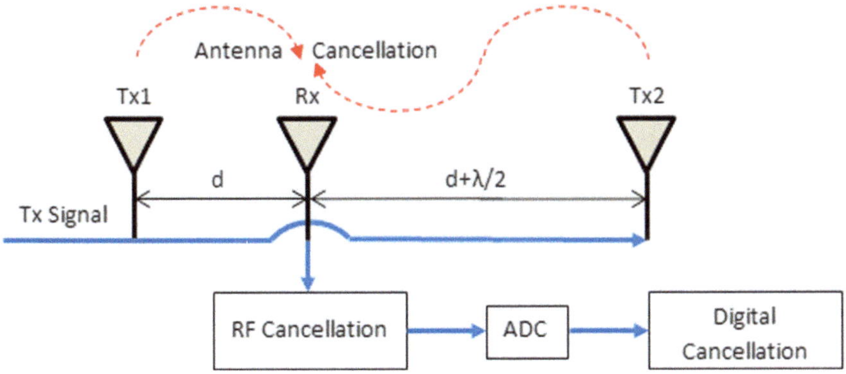

Fig. 8.2 Interference cancelation in full duplex

systems can increase the throughput three times (using 3 Tx and 3 Rx). Also, two transmit antennas can result in destructive interference in the far field.

2. Antenna cancelation does not work if the signal bandwidth is very large such as in mmWave.
3. Antenna cancelation requires manual tuning which is difficult to implement in the field.

In [3] a Balun-based architecture is proposed which solves all the above problems.

8.3 Reconfigurable Intelligent Surfaces Explained

Wireless channel is inherently unpredictable, and this results in loss of information as it travels from the transmitter to the receiver. The main reason for this is that multiple copies of the wireless signal arrive at the receiver, which sometimes add constructively and at other times destructively, causing deep fades. The deciding factor between signal copies (think of them as echoes) adding constructively or destructively is the relative phase. If the phases are aligned the signals add up but if the phases are not aligned, we get a fade (fades can be as deep as 60–80 dB). Wireless engineers over the years have worked around this problem by using multiple antennas also called antenna arrays.

For now, let us assume that there is no temporal variation of the wireless signal amplitude. The transmitter, the receiver and the environment are stationary. In such a case the relative phase between the signal copies arriving at the receive antenna array depends purely upon the environment geometry. It has been known for a long time that signal copies arriving at the array elements can be aligned by multiplication with complex weights after which these copies can be added together. It may come as surprise that this weighting can be done both at the transmitter and the receiver and is called beamforming. Beamforming essentially gives preference to signals arriving

Fig. 8.3 Reconfigurable intelligent surface

from a particular direction and negates signals arriving from unwanted direction. This preference for a certain direction is called directivity.

Transmit or receive beamforming has been around for a long time and it works like magic. But there is a downside to it; antenna arrays require multiple RF chains which can be a power hungry and expensive solution. What if instead of aligning the copies at the transmitter or receiver we do it while the signal is under transmission in the channel; this is achieved through a Reconfigurable Intelligent Surface (RIS), an electromagnetic surface which is much larger than the wavelength under consideration (Fig. 8.3). A simple analysis in [4] has shown that for a two-ray geometry (LOS plus ground reflected ray) RIS can make the power fall of as the second power of distance rather than the fourth power. When multiple meta-surfaces are combined to form an RIS there is an additional power gain proportional to square of the number of meta-surfaces, e.g., if an RIS with 100 elements is used we get a gain of $10\log10(100^2) = 40$ dB.

The main advantages of RIS over other techniques such as MIMO, Relaying, beamforming and Backscatter Communication are given below.

- RIS is nearly passive and, ideally, does not need any dedicated energy source.
- RIS can be viewed as a contiguous surface, and, ideally, any point can shape the wave impinging upon it.
- RIS is not affected by receiver noise, since, ideally, it does not need analog-to-digital or digital-to-analog converters and power amplifiers.
- RIS has full-band response, since, ideally, it can work at any operating frequency.
- RIS can be easily deployed, e.g., on the facades of buildings, ceilings of factories and indoor spaces, human clothing, etc.

Open Questions

After reading a few research papers and watching couple of YouTube tutorials, I am still not clear how is the RIS going to adjust the weights so that a beam is formed in the desired direction, the direction of the mobile user, that otherwise does not have a direct LOS with the base station. It seems that there has to be some feedback mechanism from the receiver to the RIS indicating the phase shifts required and a controller at the RIS that would obviously need some power to operate. So, the RIS would not be a totally passive device as claimed by most researchers. It is also not clear what would be the power source for the RIS. If its battery operated, how long would the battery last before it has to be charged or replaced? Lastly how many intelligent surfaces would be required per square km or per cell site as this would directly influence the infrastructure cost.

Notes:

1. At a frequency of 30 GHz, i.e., at the lower end of the millimeter wave band, the wavelength is 1 cm. A reasonably sized meta-surface (one element of an RIS) could have dimensions of 10 cm $\times$ 10 cm resulting in a surface area of 100 cm^2. An RIS with 100 of such meta-surfaces would have an area of $1m^2$, which seems practical.
2. According to published research results the RIS works best when it is placed close to the transmitter or receiver. When placed close to the base station this technique is also called holographic beamforming.
3. The usefulness of multipath has been known for a long time. In his Cellular Communication class of Fall 2000, at Virginia Tech, Dr. Ted Rappaport informed the class that the CDMA Rake Receiver likes multipath so much that sometimes artificial reflectors are put around in the environment to create multipath when there is none. That might have been the first intelligent surface to be ever used in Cellular Communications.

8.4 Index Modulation Explained

Wireless researchers are continuously exploring ways to increase the spectral efficiency (bits/sec/Hz) and energy efficiency (bits/Joule) of wireless communication systems [5]. Spectral efficiency can generally be improved by using larger constellations or by using multiple antennas at the transmitter and receiver, better known as MIMO. But increasing energy efficiency is not that straightforward. Let's consider this in bit more detail.

If the constellation size is increased, we may be able to send more bits through the channel for a given bandwidth but the bit error rate (BER) would deteriorate as minimum distance between constellation points decreases. To get the same BER we would have to bump up the transmit power (or energy). Let's look at the example below. Let's assume that we move from QPSK (2 bits per symbol) to 16-QAM (4 bits per symbol), i.e., we double the data rate. In an AWGN environment we would need 4 dB of additional energy to achieve the same BER (similar results for fading environment). So, we have not gained anything in terms of energy efficiency (Fig. 8.4).

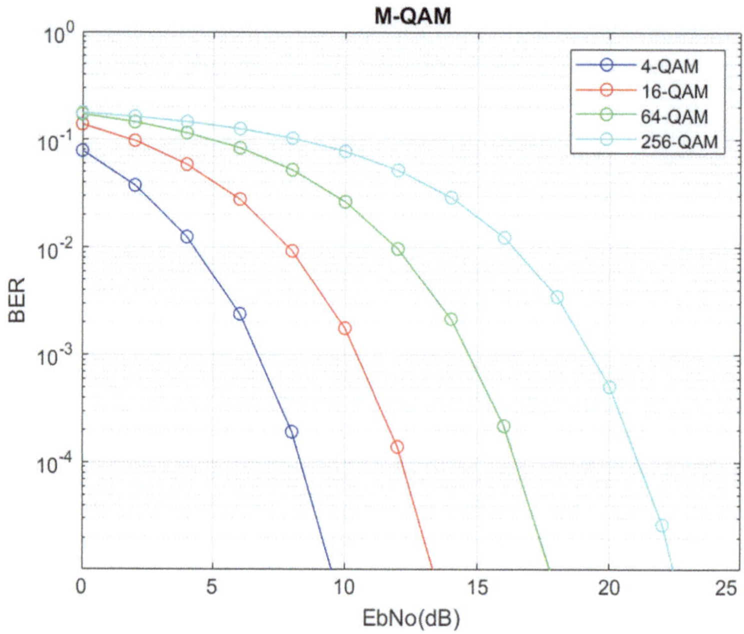

Fig. 8.4 M-QAM bit error rate in AWGN

Now let's consider the case of multiple antennas. Multiple antennas at the transmitter or receiver result in multiple RF chains, which are power hungry. In particular, simultaneous transmission from multiple transmit antennas means more power consumption [5]. In addition to power consumption in the RF front end there might also be higher power consumption at the baseband at the receiver side since more complex signal processing is required to separate the signals. Specifically, brute force methods such as maximum likelihood (ML) are quite power hungry.

A relatively new technique to improve the spectral efficiency and the energy efficiency of wireless communication system is called Index Modulation (IM) [6]. There are two main types of IM, one that uses multiple transmit antennas (Spatial Modulation) and one that uses multiple carriers (OFDM-IM). We will focus here on Spatial Modulation also known as SM.

As previously discussed, having multiple RF chains at the transmitter increases the energy consumption. What if we can use single RF chain with multiple transmit antennas, i.e., we use multiplexing. This has two advantages. An obvious one is that energy consumption is reduced at the transmitter end. A not so obvious one is that, by selecting which antenna to transmit from, there is implicit exchange of information between transmitter and receiver.

Let's assume that we have four antennas and we transmit from only one during a symbol period then we can transmit two bits by this simple selection scheme, assuming that receiver can detect that which transmit antenna was used. The total number of bits transmitted is then given as $\log_2 n_T + \log_2 M$, where n_T is the number of transmit antennas and M is the constellation size. A slightly more complex variant of this scheme is where we select a subset of transmit antennas for each symbol duration. This has the potential to further improve the spectral efficiency of the system at the cost of energy efficiency.

An Illustrative Example

Let's assume that we have BPSK modulation, and there are two transmit antennas. For each symbol period we either transmit a $+1$ or -1 from antenna Tx-1 or Tx-2. The channel gain from Tx-1 to the receiver is 0.50 and channel gain from Tx-2 to the receiver is 0.25 (we assume these gains are static and known at the receiver). So, during any given symbol period the received signal could be -0.50, -0.25, 0.25 or 0.50. That is, we have expanded the signal constellation from 2 to 4. Let's assume that we receive -0.50 at the receiver. This means that a -1 was transmitted from Tx-1. Similarly, if we receive -0.25 it means that a -1 was transmitted from Tx-2 and so on. Tx-1 or Tx-2 or for that matter any number of antennas could be selected based upon a lookup table at the transmitter side. Tx-1 could be selected when the input is -1 (or binary 0) and Tx-2 could be selected when input is $+1$ (or binary 1).

Applying the formula, we discussed above, number of transmitted bits can be calculated as

$$\log_2 n_T + \log_2 M = \log_2 2 + \log_2 2 = 1 + 1 = 2 \text{ bits/time slot}$$

A last point about antenna correlation. Assume that the transmit antennas are closely spaced and the channel coefficients are 0.25 and 0.30, respectively. Then the expanded constellation points would be much closer together and the probability of error increases!

8.5 Ray-Tracing for Network Planning

It's very easy to get lost in the jargon when selecting a simulation tool for planning your wireless network. You will be faced with complex terminology which would not make much sense. At one end of the spectrum are solutions based on simple empirical models while at the other end are solutions based on ray-tracing techniques [7]. Empirical models are based on measurement data and are your best bet if you want a quick and cheap solution, whereas ray-tracing techniques are based on laws of physics and promise more accurate results. In principle ray-tracing techniques are quite simple: Just transmit a bunch of rays in all directions and see how they behave. However, when the number of rays and their interactions becomes large the simulation may become prohibitively complex. The simulation time for complex geometries may vary from a few hours to several days. Following are some of the factors that you must consider when selecting a ray-tracing simulator.

1. **Upper limit on the number of interactions**

Ray-tracing simulators essentially generate a bunch of rays (image-based techniques are an exception) and then follow them around as they reflect, refract, diffract and scatter. Each interaction decreases the strength of the rays. The strength of the rays also decays with distance (depends!). As a result, the simulator needs to decide when to terminate a ray path. This is usually done based upon the number of interactions that a ray undergoes (typically 8–10 interactions are considered) or based upon its strength (once the strength of a ray falls below − 110 dBm there is no point following it any further). Higher the number of interactions considered, greater the accuracy of the simulation but higher the computational complexity.

2. **Granularity in field calculations**

Field calculations cannot be performed at each and every point within the simulation space. The usual approach is to divide the region under study into a grid such that locations closer to a transmitter are covered more finely and the regions further away are covered in lesser detail. The rays are then combined within each block of the grid to get the resultant field strength. The level of granularity determines the computation load. It would be prohibitively expensive to have a very high level of granularity for a large network.

3. **Accuracy in modeling the various propagation phenomenon**

As mentioned previously an accurate modeling of all propagation phenomena is required including reflection, refraction, diffraction and scattering. Some ray-tracing simulators might model reflection and refraction only while ignoring the other phenomenon such as diffraction. Furthermore, some ray-tracing simulators might consider all reflections to be specular (no scattering). This is a good approximation for large smooth surfaces but is not such a good assumption for irregular terrain.

4. **Granularity of the terrain database**

Most state-of-the-art ray-tracing tools use some sort of terrain database to perform their calculations. These terrain databases are required for determining the paths of the rays as they travel in dense urban environments and may contain simple elevation data or actual 3D building data. These databases may have accuracy of 10 or 30 m or maybe more. The accuracy of the simulation is highly dependent on the granularity of the terrain database.

5. **Accuracy in representation of building materials**

The wireless signal propagation within cities is governed by complex phenomena such as reflection, refraction, diffraction and scattering. Let's take the example of the phenomenon of reflection. The percentage of signal reflected back at a particular interface is dependent on permittivity and permeability of the object. Based on these properties only 10% of the signal maybe reflected or 50% of the signal may be reflected. So, for accurate simulation not only should we have a high level of granularity of the 3D building data, we also need an accurate description of the building materials.

6. **Dynamic Channel Behavior**

A wireless channel is continuously changing, i.e., the channel is dynamic (as opposed to being static). However, the ray-tracing techniques available in the literature do not capture this dynamic behavior. The dynamic behavior of the channel is mainly due to the motion of the transmitter or receiver as well as motion of the surroundings. While the position of the transmitter and receiver can be varied in the ray-tracing simulation the surroundings are always stationary. Hence a ray-tracing simulator is unable to capture the time-varying behavior of the channel.

The accuracy of ray-tracing simulators is bound to increase as the computational power of computers increases and as accurate 3D building databases become available throughout the world. Until that time, we would have to fall back to approximate simulations or maybe measurement results.

8.6 60 GHz Millimeter Wave Band—Seems Like a Free Lunch

Let us start by first listing down the advantages of the 60 GHz Millimeter Wave Band, a band spread between 57 and 64 GHz. This unlicensed band was first released in the USA in 2001 but with limited allowance for transmit power (EIRP of 40 dBm). Later on, in 2013, this limit was increased to allow for greater transmit power (EIRP of 82 dBm) and larger range. The higher EIRP can be achieved with an antenna gain of 51 dBi or higher (EIRP is simply the product of transmit power and antenna gain). But first the advantages:

1. Unlicensed band means you do not have to pay for using the frequencies in this band.
2. Wide bandwidth of 7 GHz allows high data rate transmissions. Remember Shannon Capacity Theorem?
3. High atmospheric absorption (Fig. 8.5) resulting in greater path loss (up to 20 dB/km) and shorter range. This means lesser cochannel interference and higher reuse factor.
4. Smaller antenna sizes allowing for multiple antennas to be put together in the form of an array providing high gain.
5. This band is quite mature and electronic components are cheap and easily available.

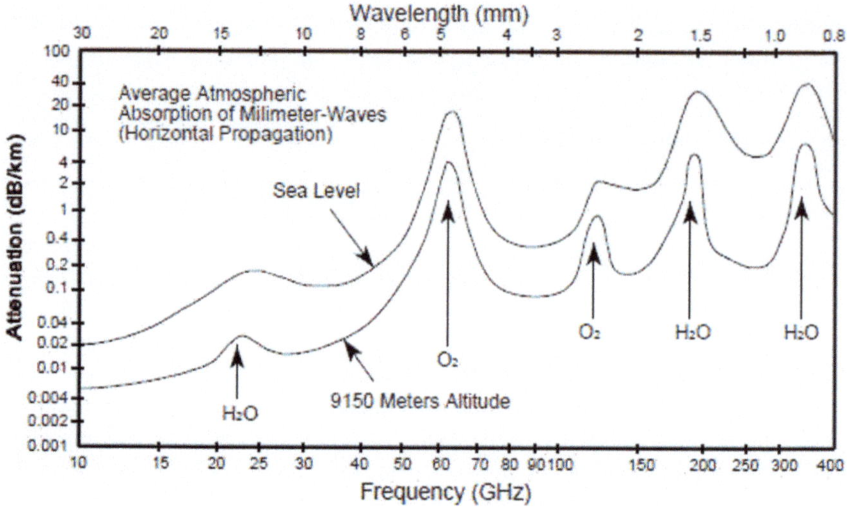

Fig. 8.5 Atmospheric absorption at millimeter wave [8]

While most of the above points make sense, points (3) and (4) need a bit more clarification.

Higher Path Loss

At first higher path loss and shorter range seems to be a disadvantage. Moreover, if we want a shorter range, we do not need to increase the path loss we can just decrease the transmit power. After all, Receive Power = Transmit Power-Path Loss. But to understand this fully we need to look at the receive power vs distance curve shown below. Higher path loss exponent (rate at which the received power falls off with distance) means that the slope of the line is higher. So, at shorter distance the received power would be higher giving higher signal to noise ratio (SNR) and greater Shannon Capacity and/or lower bit error rate (BER). While at larger distances the received power would fall off sharply resulting in lower received power and lower cochannel interference, allowing for a higher frequency reuse factor (Fig. 8.6).

Higher Antenna Gain

The energy collected by an antenna depends on its aperture or simply its physical size. The antenna aperture can be mathematically written as

$$A = G\lambda^2/4\pi$$

So, with increasing frequency or decreasing wavelength the antenna aperture decreases. But what about antenna gain? Gain of an antenna is independent of frequency and remains constant. For example, a half-wave dipole has a gain 1.64 (2.15dBi) irrespective of the frequency of operation. However, the gain can be increased by using multiple antennas of relatively small gain each. If we use two

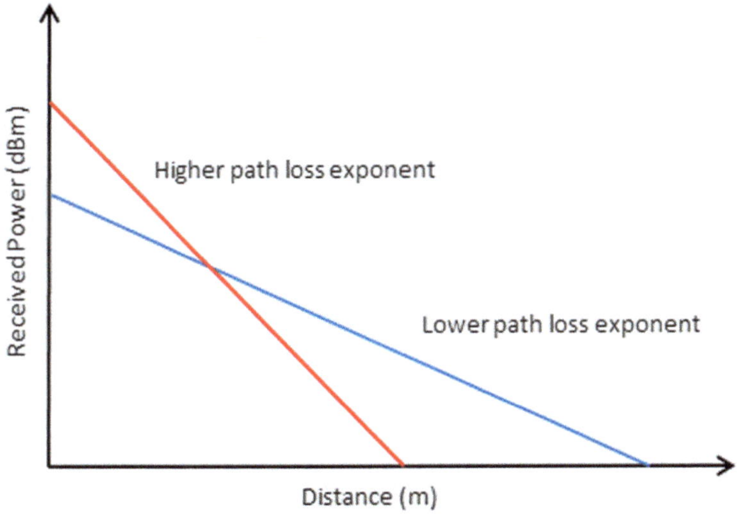

Fig. 8.6 Received power as a function of distance

antenna elements the gain increases by 3 dB and similarly for four antenna elements the gain increases by 6 dB. At 60 GHz the wavelength is only 0.5 cm, consequently the antenna size is also quite small, and a large antenna array comprising possibly hundreds of elements can be put together to achieve high gain and high aperture.

Note:

1. Received power is usually proportional to $1/d^n$ where d is the transmit receive separation and n is the path loss exponent. The value of n is two in free space line of sight communication. But in dense urban environments it could be four or higher. The straight-line behavior shown above is only observed when received power is in dBm and x-axis is on logarithmic scale.
2. A word about Shannon Capacity. When the bandwidth increases there are two factors at play. First of all, the capacity increases with increase in bandwidth but noise power ($P_n = kTB$) also increases due to which capacity decreases. The first factor is so overpowering that the combined response is almost a linear increase in C with an increase in B.
3. Let us assume that the received signal power P_s is – 70 dBm and bandwidth B is 7 GHz. This gives us a noise power P_n of – 75.4 dBm and an SNR of 5.4 dB. The Shannon Capacity is thus calculated as 15.13 Gbps. This simple example just demonstrates the power of having a wideband channel like we have at 60 GHz.
4. After revision, the 60 GHz band covers frequencies from 57 to 71 GHz (14 GHz of bandwidth). The frequency band is subdivided into 6 different channels in IEEE 802.11ad with each channel occupying 2160 MHz of BW.

8.7 5G Millimeter Waves—Are They Really Harmful?

There has been a continuous debate about harmful effects of electromagnetic radiations ever since they came into existence. Most of the research results suggest that there are no harmful effects, if the rules and regulations are followed. But there is a small body of research that suggests that there might be some harmful effects and more research needs to be carried out. This is particularly important now as 5G Wireless Technology is being rolled out around the world [9], and it uses millimeter waves for which we have limited data. Also, 5G would be using much smaller cells meaning that base stations would be closer to human beings.

Those who believe that EM radiation in the millimeter wave is harmless argue that the waves in this band are of the non-ionizing type, and the only effect they have on the human body is some localized heating [10–13, 15, 16]. By non-ionizing it is meant that the photon energy in this band is so low that it cannot knock out an electron from an atom or molecule. Just to emphasize this further it must be mentioned that a photon in this band has energy of 1.2 meV, whereas 12 eV is required to remove an electron from its parent body, i.e., the energy of the photon is 10,000 times less than the minimum required [15]. Now let us look at the factors that need to be considered when assessing 5G technology and its harmful effects, if any (there are some studies that raise red flags like [17]).

1. **Transmit Power**

The transmit power of a 5G base station can vary from 250mW to 120W depending upon the size of the cell. Compared to 2G/3G/4G towers which could have a maximum transmit power of 20W, this is about six times higher. On a logarithmic scale its 7.8 dB higher. But radiation from base stations is not such a big concern since power falls of as the squared of the distance in free space and as fourth power of distance in an urban environment. The more important metric to look at is transmit power of the mobile station. We know that a GSM (2G) mobile station could have a transmit power of up to 1000 mW. Compared to this a 5G cell phone has a maximum transmit power of only 200 mW, 7 dB lower.

2. **Cell Size and Base Station Antenna Height**

As mentioned earlier the 5G cell sizes are expected to be much smaller [13], particularly in dense urban environments. A picocell can have a radius of 100–200 m, whereas an indoor femtocell can have a radius of 10 m or even lesser. Furthermore the height of the base station antenna is going to be much lower as antennas are to be deployed on lamp posts, bus shelters, etc. 3G networks reached densities of 4–5 base stations per squared km, 4G networks reached densities of 8–10 for the same area, while 5G networks could achieve densities of 40–50. So exposure to radiation would definitely be higher.

3. **Massive MIMO and Beamforming**

It is well known that 3G and 4G systems use MIMO technology to get better spectral efficiency, reliability and capacity. 5G systems take this to whole new level by employing 64 × 64 antenna configurations. Using these antennas a base station can form a beam on a user, i.e., it will transmit more power in one direction and transmit lesser or no power at all in the other directions. This means that a base station can reuse the same time slot and frequency in another direction (remember the concept of frequency reuse). But what about exposure to higher powers in the vicinity of the base station? The narrower a beam gets the more power density it would have (PD is given in Watts per square meter). Since the base station would be lower, an unintended signal recipient in the direction of the beam might get exposed to higher power density.

4. **Power Control**

The strength of a wireless communication signal varies greatly as it proceeds from the transmitter to the receiver. There are two main components to this, a distance dependent path loss and fading which occurs due to constructive and destructive interference of the wireless signal. To overcome these effects the transmitter adjusts it's transmit power so that a good quality wireless link can be maintained. When the cell size is small, as in 5G (usually), this is easier to implement. But if the cell size is large and user is on the cell edge the transmit power has to be substantially increased which can cause cochannel interference and could be harmful to a cell phone user as well. Last point to note is that most wireless communication systems, including 5G, use Adaptive Modulation and Coding Schemes (MCS) and to achieve a high throughput higher power is transmitted than is necessary just to maintain the link.

5. **Near Field and Far Field**

Electromagnetic radiation from an antenna can be divided into three regions based upon the distance of the observer from the antenna: a reactive near field, radiative near field and radiative far field. Most of the analysis in the literature assumes that we are in the far field of the antenna where the Electric field, magnetic field and direction of propagation, are all perpendicular to each other and ratio of electric to magnetic field is a constant. But the near field which stretches to about half the wavelength from the antenna (dipole) is not that well understood. One thing that is known is that the electric field and magnetic field fall off much more rapidly in the near field than in the far field. At 1 GHz the near field is within 15 cm of the antenna whereas at 30 GHz this is reduced to half a cm. So for millimeter waves we can say that a mobile phone user is in the far field of the mobile antenna most of the time, which is studied in detail and well understood.

6. **Penetration in the Human Body**

According to studies conducted in the millimeter wave band about 30%–40% of the EM energy is reflected back by the human skin. But the energy reflected back decreases with increasing frequency, e.g., at 40 GHz, 43% of the energy falling normally on the human body is reflected back while at 100 GHz this is reduced to 30% only. On the other hand penetration loss within the human body increases with increasing frequency. It is reported that 90% of the millimeter wave energy is absorbed within the first two layers of the skin, namely epidermis and dermis, which are only a few millimeters in thickness [12].

7. **Superposition of Signals**

Most of the studies on effects of exposure to radiation consider one type of radiation, at a particular power level and for a limited time. But real-life scenarios are quite different. A typical human being is subjected to a variety of signals at any instant. This may include Bluetooth, Wi-Fi, UWB, 3G/4G/5G, radiation from microwave ovens, etc. With the advent of Internet of Things (IoT) it's not uncommon for a typical home to have tens of devices, in closed proximity, transmitting simultaneously. So one signal might not be harmful for a human being but when so many signals are combined what effect do they have? Are there any studies that have observed the impact of these devices when operated in close proximity for a number of years? Are results on animals directly applicable to humans? These are some of the questions that need to be answered before adopting this new technology.

Appendix: Measurement of Radiation

Specific Absorption Rate (SAR)

SAR is the most common parameter used to measure amount of EM radiation that a body is exposed to but it is only useful if you have a volume under consideration. In the USA, FCC requires that phones sold have a SAR level at or below 1.6 watts/kg averaged over 1 g of tissue. The ICNIRP SAR limit for mobile devices is 2 watts/kg averaged over 10 g of tissue. For long-term exposure the limits are much more stringent, particularly for the general public.

Power Density (PD)

If you do not have a defined volume but have a surface area to work with then power density is the most useful measure. The PD limit used most often is 1mW per square centimeter [14, 15] (sometimes also given in watts per square meter).

Temperature Increase

The most direct impact of non-ionizing radiation is an increase in temperature; therefore, it may also be used to measure the exposure to EM radiation. The limit for this is 1 degree centigrade increase in temperature for the body part exposed to radiation. This is a useful measure in the near field of a radiating antenna where the EM energy may be difficult to measure.

Federal Communication Commission (FCC) Advice for General Public

As per information given on FCC website [18] "*For users who are concerned with the adequacy of this standard (FCC defined SAR limits) or who otherwise wish to further reduce their exposure, the most effective means to reduce exposure are to hold the cell phone away from the head or body and to use a speakerphone or hands-free accessory. These measures will generally have much more impact on RF energy absorption than the small difference in SAR between individual cell phones.*"

8.8 Soft Frequency Reuse

Frequency Reuse is a well-known concept that has been applied to wireless systems over the past three decades, e.g., in GSM systems. As the name suggests Frequency Reuse implies using the same frequencies over different geographical areas. If we have a 25 MHz band then we can have 125 GSM channels and 125 * 8 = 1000 time multiplexed users in a given geographical area. Now if we want to increase the number of users we would have to reuse the same frequency band in a geographically separated area. The technique usually adopted is to use a fraction of the total frequency band in each cell such that no two neighbor cells use the same frequency. Typically the frequency band is divided into 3 or 7 cells.

The division of the frequency band in to smaller chunks reduces the system capacity, e.g., one cell with 25 MHz bandwidth would have much higher capacity than 7 cells having 3.5 MHz each. To overcome this problem a frequency reuse of 1 has been proposed, i.e., each cell has the full system bandwidth (nearly). The problem of cochannel interference at the cell boundaries is resolved by dedicating a small chunk of the available spectrum for the cell edges (Fig. 8.7).

In Soft Frequency Reuse (SFR) the cell area is divided into two regions; a central region where the entire frequency band is available and a cell edge area where only a small fraction of the spectrum is available [19]. The spectrum dedicated for the cell edge may also be used in the central region if it is not being used at the cell edge. The lack of spectrum at the cell edge may result in much reduced Shannon Capacity for that region. This is overcome by allocating high power carriers to the users in this region thus improving the SINR and the Shannon Capacity.

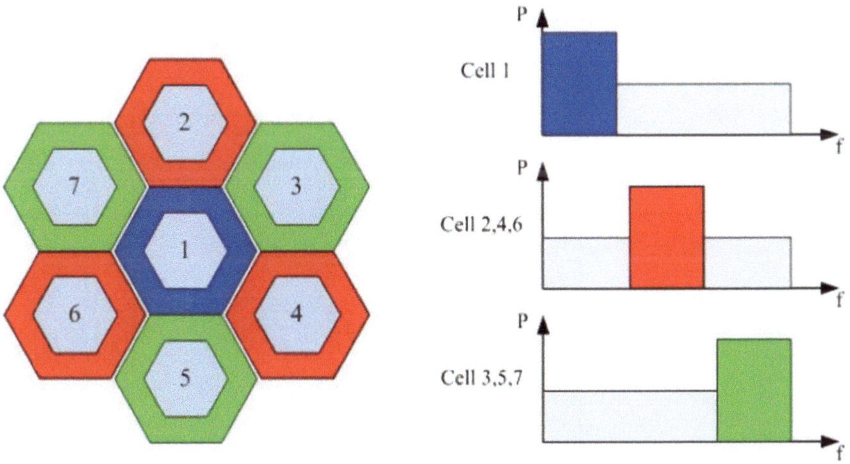

Fig. 8.7 Frequency planning and power allocation for SFR scheme

Note:

1. The Signal to Interference and Noise Ratio is given as:
   SINR = Signal Power/(Intercell Interference + Intracell Interference + AWGN Noise)
2. Typically the term capacity was used to describe the number of voice channels (or users) that a system can support. But with modern digital communication systems it usually refers to the Shannon Capacity that can be achieved (in bits/sec/Hz).

8.9 Patch Antenna Design Using Transmission Line Model

A microstrip patch antenna can be designed using either the transmission line model or the cavity model (more complex models also exist that suit a particular design). We here demonstrate the transmission line model since it is fairly simple to implement and results in antenna designs with reasonably good performance in terms of return loss and efficiency [20] (Fig. 8.8).

The design starts with selecting the operating frequency, selecting a substrate with the required permittivity and defining the height of the substrate. Thick substrates with low permittivity result in antenna designs with high efficiency and large bandwidths. Thin substrates with high permittivity lead to a smaller antenna size but with a lower bandwidth and a high-radiation loss. The trade-offs between substrate thickness and permittivity and antenna bandwidth and efficiency have been widely discussed in the literature.

According to the transmission line model, the length L and width W of the patch are calculated as

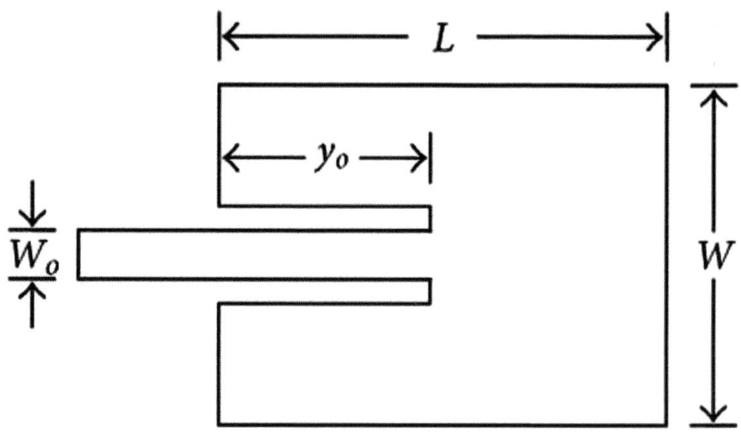

Fig. 8.8 Patch antenna construction

$$W = \frac{v_o}{2f_r}\sqrt{\frac{2}{\varepsilon_r + 1}}$$

$$L = \frac{v_o}{2f_r\sqrt{\varepsilon_{\mathrm{reff}}}} - 2\Delta L$$

$$\varepsilon_{\mathrm{reff}} = \frac{\varepsilon_r + 1}{2} + \frac{\varepsilon_r - 1}{2}\left[1 + 12\frac{h}{W}\right]^{-1/2}$$

$$\Delta L = 0.412h\frac{(\varepsilon_{\mathrm{reff}} + 0.3)(W/h + 0.264)}{(\varepsilon_{\mathrm{reff}} - 0.258)(W/h + 0.8)}$$

where v_o is the speed of light in free space, f_r is the resonant frequency, ε_r is the relative permittivity, ΔL is the extension in length due to fringing effect and h is the height of the substrate.

Although the design of the patch is quite simple, the design of the feeding mechanism is not that straight forward. It is the experience of the author that the design of feeding mechanism needs a bit of trial and error and a simulation software such as CST can be very useful for this purpose. There are four possible methods that can be used:

i. Microstrip-line feed
ii. Probe feed
iii. Aperture-coupled feed
iv. Proximity-coupled feed

Now a little bit about the behavior of the patch antenna once the design and fabrication is complete. The electric and magnetic field variations within a patch are sometimes a bit confusing and difficult to visualize. The figure below shows the E and H field variations within a rectangular patch of length L and width W (Fig. 8.9).

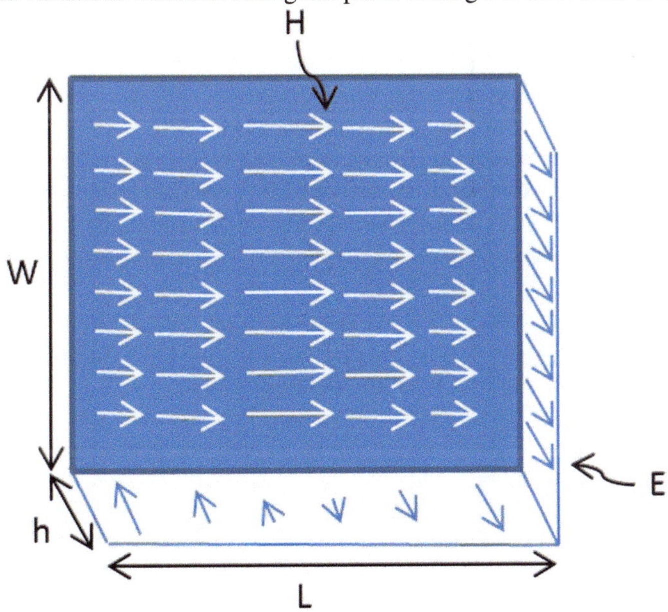

Fig. 8.9 E and H field of a patch antenna

As can be seen the E-field varies along the length of the patch with minimum at the center and maximum at the edges (maximum positive and maximum negative). The H-field also varies along the length in a direction perpendicular to the E-field. The H-field is maximum at the center and minimum at the edges. Thus the impedance is zero at the center of the patch (remember Ohms law, $Z = V/I$).

Another point to note is that the E-field does not completely terminate at the edges along the length of the antenna rather it extends at the outer periphery. These field extensions are known as fringing fields and cause the patch to radiate [20].

Questions and Numerical Problems

1. Compare the size of an 8×8 rectangular antenna array composed of dipole elements at (a) 1 GHz (b) 5 GHz (c) 30 GHz. You can assume that interelement spacing is half the wavelength.
2. What is the maximum gain that can be achieved by using an (a) 64 element rectangular antenna array (b) 64 element circular antenna array? Plot the 3D radiation patterns.
3. Calculate the maximum gain of a concentric circular array for the following array sizes (a) 16 elements (b) 32 elements (c) 64 elements. Plot the 3D radiation patterns.
4. Explain the concept of interference cancelation in Full Duplex mode using multiple antenna elements? What problems do you foresee in its practical implementation?
5. Discuss the pros and cons of using a passive relay instead of active relay.
6. Simulate the bit error rate performance of a 4×1 MIMO system using Index Modulation. Comment on the improvement (if any) in spectral efficiency and energy efficiency.
7. Simulate the received power as a function of distance using the 2-Ray model. You can assume perfect ground reflection and unity antenna gains. Other parameters such as transmit power, frequency and antenna heights may be assumed as appropriate.
8. Is it true that path loss increases with frequency? What does Friis transmission equation say about it? What can you deduce from it?
9. What is the typical path loss exponent in (a) Rural areas (b) Urban areas? Is a higher path loss exponent necessarily bad for wireless network operation?
10. Explain the concept of atmospheric absorption of electromagnetic waves. Does it have any substantial impact on the performance of wireless communication systems?
11. Please write a detailed review of the article about the harmful effects of 5G radiation? Is localized heating the only adverse effect of wireless radiation (the student is advised to do a literature survey for this).
12. Compare the capacity of two wireless networks; one with a spectrum allocation of 20 MHz per cell (frequency reuse of 1) and one where the 20 MHz spectrum is divided among 4 cells (frequency reuse of 4). Assume that both the configurations cover the same geographical area and a single channel occupies 1 MHz.

13. Explain the concepts of cochannel interference and adjacent channel interference in wireless communications. What is the concept of guard band and what is the typical size of a guard band?
14. What is the phenomenon that results in a patch antenna radiating energy in the free space?
15. What are the most important factors that must be considered when acquiring a ray-tracing simulator for wireless network planning? Please elaborate.

Useful Links

1. Beyond Massive MIMO
   https://www.raymaps.com/index.php/beyond-massive-mimo/
2. Reconfigurable Intelligent Surfaces Explained
   https://www.raymaps.com/index.php/reconfigurable-intelligent-surfaces-explained/
3. Index Modulation Explained
   https://www.raymaps.com/index.php/index-modulation-explained/
4. Ray-Tracing for Network Planning
   https://www.raymaps.com/index.php/ray-tracing-for-network-planning-ii/
5. 60 GHz Millimeter Wave Band—Seems Like a Free Lunch
   https://www.raymaps.com/index.php/60-ghz-millimeter-wave-band-seems-like-a-free-lunch/
6. 5G Millimeter Waves—Are They Really Harmful?
   https://www.raymaps.com/index.php/5g-millimeter-waves-are-they-really-harmful/
7. Soft Frequency Reuse
   https://www.raymaps.com/index.php/soft-frequency-reuse/
8. Patch Antenna Design Using Transmission Line Model
   https://www.raymaps.com/index.php/patch-antenna-design-using-transmission-line-model/

References

1. Bjornson, E., Sanguinetti, L., Wymeersch, H., Hoydis, J., Marzetta, T.L.: Massive MIMO is a reality—what is next? Five promising research directions for antenna arrays. arXiv e-prints, p. arXiv:1902.07678 (2019)
2. Fishler, E., Haimovich, A., Blum, R., Chizhik, D., Cimini, L., Valenzuela, R.: MIMO radar: an idea whose time has come. In: IEEE Radar Conference, pp. 71–78 (2004)
3. Jain, M., Choi, J., Kim, T., Bharadia, D., Seth, S., Srinivasan, K., Levis, P., Katti, S., Sinha, P.: Practical, real-time, full duplex wireless. In: MobiCom '11: proceedings of the 17th annual international conference on mobile computing and networking, pp. 301–312 (2011)

4. Basar, E., Di Renzo, M., de Rosny, J., Debbah, M., Alouini, M.S., Zhang, R.: Wireless communications through reconfigurable intelligent surfaces. arXiv eess.SP
5. Jiang, J., Dianati, M., Imran, M.A., Tafazolli, R., Chen, Y.: On the Relation between energy efficiency and spectral efficiency of multiple-antenna systems. IEEE Trans Veh Technol (2013)
6. Basar, E.: Index modulation techniques for 5G wireless networks. IEEE Commun. Mag. (2016)
7. Yun, Z., Iskander, M.F.: Ray tracing for radio propagation modeling: principles and applications. IEEE Access **3**, 1089–1100 (2015)
8. https://www.everythingrf.com/community/what-is-the-impact-of-the-earths-atmosphere-on-rf-signal-propagation
9. https://www.qualcomm.com/5g/what-is-5g
10. https://www.nature.com/articles/d42473-019-00009-7
11. https://www.nature.com/articles/s41370-021-00297-6
12. https://www.ncbi.nlm.nih.gov/pmc/articles/PMC4629874/pdf/nihms-695510.pdf
13. https://spectrum.ieee.org/will-5g-be-bad-for-our-health
14. https://futurenetworks.ieee.org/tech-focus/september-2019/5g-communications-systems-and-radiofrequency-exposure-limits
15. https://www.infineon.com/dgdl/Infineon-Health%20Effects%20of%20mmWave%20Radiation-PI-v01_01-EN.pdf?fileId=5546d46266a498f50166f1ada0520444
16. https://www.cnet.com/tech/mobile/is-5g-making-you-sick-probably-not/
17. https://www.nytimes.com/2019/07/16/science/5g-cellphones-wireless-cancer.html
18. https://www.fcc.gov/engineering-technology/electromagnetic-compatibility-division/radio-frequency-safety/faq/rf-safety
19. Yu, Y., Dutkiewicz, E., Huang, E., Mueck, M., Fang, G.: Performance analysis of soft frequency reuse for inter-cell interference coordination in LTE Networks. ISCIT (2010)
20. Ahmed, Y., Hao, Y., Parini, C.: A 31.5 GHz patch antenna design for medical implants. Int. J. Antennas Propag. **2008**, 6, Article ID 167980 (2008)

Chapter 9
Simulation in Python

9.1 Introduction

This is the last chapter of this book, but an old debate, proprietary software versus open-source software. Most of the engineers of my generation have grown up using MATLAB for their design and simulation tasks. We have become so comfortable with it that we now even think in terms of vectors and matrices. We apply an averaging filter all the time in our daily life, taking out the inherent randomness. But there are alternatives now, like an open source version of MATLAB, by the name of Octave. Then there is Python which also is also free and open source with tons of libraries and developer resources. So let's take a deeper dive to discuss the pros and cons of each.

Let us start with the advantages of Python over MATLAB. First of all Python is free and open-source, whereas MATLAB is quite expensive, especially if you are a large company which requires hundreds of concurrent licenses. Secondly, it has a large developer community which is continuously contributing to the ever increasing number of libraries. While MATLAB has only one Integrated Development Environment (IDE), Python users can choose their IDE from very simple ones to more advanced ones with a lot of features. Python is, just like MATLAB, a cross-platform language which can run on all Operating Systems—even embedded systems having a small Linux kernel. All major AI/ML frameworks are based on Python such as Tensorflow, Keras, PyTorch. Lastly, the strength of Python can be gauged from the fact that the image processing for M-87 Black Hole was done using Python (https://numpy.org/case-studies/blackhole-image/).

Now we move over to the advantages of MATLAB. MATLAB is supported by Mathworks, a company which has an employee base of 5000 and \$1.25 billion in revenue. It will not disappear in thin air one day and shut down its support of the software that you have purchased and spent the last ten years learning. Once you install MATLAB along with the toolboxes, you do not have to worry about what's included and what's not. Although this point is debatable, I have found that MATLAB

Y. Ahmed, *Recipes for Communication and Signal Processing*,
Signals and Communication Technology,
https://doi.org/10.1007/978-981-99-2917-7_9

code generally executes faster than Python code on my machine. Another great thing about MATLAB is the graphical programming tool called Simulink which is used for modeling, simulating and analyzing multidomain dynamical systems. Lastly, if MATLAB is being used by everybody in your community, it's easier to discuss problems and their solutions.

In this chapter we have shown three bit error rate calculations; binary phase shift keying in AWGN, Alamouti code performance in a fading channel and OFDM performance in a fading channel. We also have shown how to generate time-varying correlated Rayleigh fading envelope in Python. Lastly, we have done a performance comparison of MATLAB and Python in terms of execution speed. The performance of MATLAB is better for random variable generation, for "for loop" implementation and for comparing two vectors. MATLAB is also faster for calculation of bit error rate of BPSK for 10 values of SNR. The only area where Python has superior performance is in plotting a histogram and a scatter plot. But matplotlib offers limited functionality as compared to plot function in MATLAB. Overall the experience of moving from MATLAB to Python was quite good and the learning curve was not that steep.

The author found online forums such as those maintained by the Python community and others such as Stackoverflow and Quora to be quite useful.

9.2 BPSK Bit Error Rate Calculation Using Python

Have you ever thought about how life would be without MATLAB? As it turns out there are free and open source options such as Python. We have so far restricted ourselves to MATLAB in this book but now we venture out to find out what are the other options. Given below is a most basic Python code that calculates the bit error rate of binary phase shift keying (BPSK). Compare this to our MATLAB implementation earlier in Chap. 2.

There are various IDEs available for writing your code but I have used Enthought Canopy Editor (32 bit) which is free to download and is also quite easy to use [enthought.com]. So, as it turns out that there is life beyond MATLAB. In fact, there are several advantages of using Python over MATLAB which we will discuss later in another post. Lastly, please note the indentation in the code below as there is no "end" statement in a "for loop" in Python (Fig. 9.1).

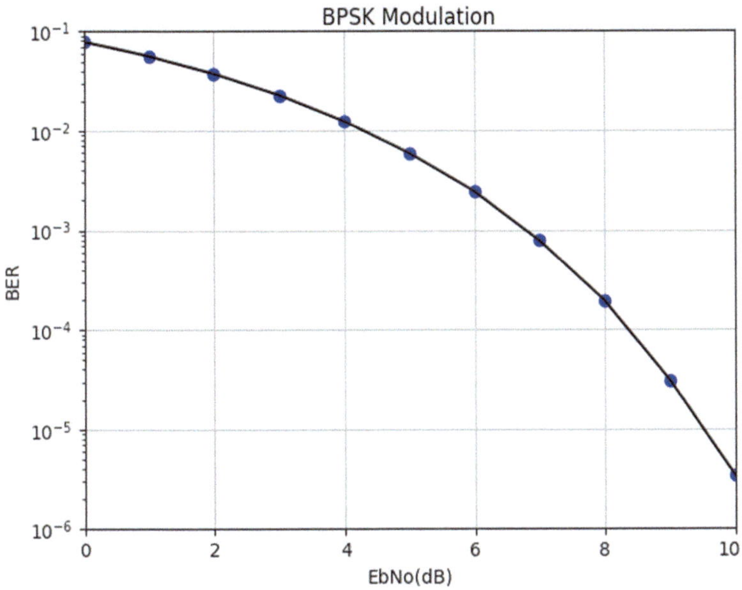

Fig. 9.1 BPSK bit error rate

```
##########################################################################################
#                              BIT ERROR RATE OF BPSK
#           N is the number of bits or symbols passed through the channel
#                   EbNo is the energy per bit to noise PSD ratio
#                          ber is the output bit error rate
#                           Copyright 2020 RAYmaps
##########################################################################################
from numpy import sqrt
from numpy.random import rand, randn
import matplotlib.pyplot as plt

N = 5000000
EbNodB_range = range(0,11)
itr = len(EbNodB_range)
ber = [None]*itr
```

```
for n in range (0, itr):
    EbNodB = EbNodB_range[n]
    EbNo=10.0**(EbNodB/10.0)
    x = 2 * (rand(N) >= 0.5) - 1
    noise_std = 1/sqrt(2*EbNo)
    y = x + noise_std * randn(N)
    y_d = 2 * (y >= 0) - 1
    errors = (x != y_d).sum()
    ber[n] = 1.0 * errors / N
    print "EbNodB:", EbNodB
    print "Error bits:", errors
    print "Error probability:", ber[n]
plt.plot(EbNodB_range, ber, 'bo', EbNodB_range, ber, 'k')
plt.axis([0, 10, 1e-6, 0.1])
plt.xscale('linear')
plt.yscale('log')
plt.xlabel('EbNo(dB)')
plt.ylabel('BER')
plt.grid(True)
plt.title('BPSK Modulation')
plt.show()
#############################################################################################
```

9.3 MATLAB Versus Python Computational Speed

Operating System

- Windows

System

- Processor Intel(R) Core(TM) i7-5500U CPU @ 2.4 GHz
- Installed Memory 8.00 GB
- System Type 64 Bit Operating System, × 64 Based Processor

Integrated Development Environment (IDE)

- Enthought Canopy
- Version 2.1.3.3542 (32 bit)

Operation	Time in sec (MATLAB)	Time in sec (PYTHON)
10 million uniform random variable generation	0.10	0.15
10 million normal random variable generation	0.13	0.40
For loop counting up to 100 million	0.40	11.60
Comparing two vectors of length 10 million each	0.39	0.55
Plotting a histogram of 10 million values	0.89	0.76
Plotting a scatter plot of 1 million values	0.30	0.23

(continued)

(continued)

Operation	Time in sec (MATLAB)	Time in sec (PYTHON)
Bit error rate calculation of BPSK for 10 values of SNR	2.49	4.51

Python is a bit slower than MATLAB for most of the cases but the real difference is in implementation of "for loop" where the speed of MATLAB is 29 × that of Python. A surprising result was that the plot functions for Python were somewhat faster than MATLAB.

9.4 Alamouti—Transmit Diversity Scheme

We have already seen in previous posts that the BER of BPSK increases significantly when the channel changes from a simple AWGN channel to a fading channel. One solution to this problem, that was proposed by Alamouti, was to use transmit diversity, i.e., multiple transmit antennas transmit the information over multiple time slots increasing the likelihood of receiving the correct information. We have considered the simplest case of two transmit antennas and BPSK modulation (QPSK modulation would give the same BER with twice the throughput). Given below is the Python code for this, feel free to modify it.

```
##########################################################################################
#                    ALAMOUTI SCHEME IMPLEMENTED IN PYTHON
#          N is the number of bits or symbols passed through the channel
#                    EbNo is the energy per bit to noise PSD ratio
#                         ber is the output bit error rate
#                          Copyright 2020 RAYmaps
##########################################################################################
from time import time
from numpy import sqrt
import random
import matplotlib.pyplot as plt

t=time()

N = 100000
EbNodB_range = range(0, 11)
itr = len(EbNodB_range)
ber = [None]*itr

for n in range (0, itr):
        EbNodB = EbNodB_range[n]
        EbNo=10.0**(EbNodB/10.0)
        noise_std = 1/sqrt(2*EbNo)
        noise_mean = 0
        no_errors = 0
```

```
        for m in range (0, N):
                tx_symbol1 = (2*random.randint(0,1)-1)
                tx_symbol2 = (2*random.randint(0,1)-1)
                noise1 = (random.gauss(noise_mean, noise_std)+
                1j*random.gauss(noise_mean, noise_std))
                noise2 = (random.gauss(noise_mean, noise_std)+
                1j*random.gauss(noise_mean, noise_std))
                ch_coeff1 = (random.gauss(0,1/sqrt(2))+
                1j*random.gauss(0,1/sqrt(2)))
                ch_coeff2 = (random.gauss(0,1/sqrt(2))+
                1j*random.gauss(0,1/sqrt(2)))
                rx_symbol1 = ((1/sqrt(2))*tx_symbol1*ch_coeff1+
                (1/sqrt(2))*tx_symbol2*ch_coeff2 + noise1)
                rx_symbol2 = (-(1/sqrt(2))*tx_symbol2*ch_coeff1+
                (1/sqrt(2))*tx_symbol1*ch_coeff2 + noise2)
                estimate1 = (ch_coeff1.conjugate()*rx_symbol1+
                ch_coeff2*rx_symbol2.conjugate())
                estimate2 = (ch_coeff2.conjugate()*rx_symbol1-
                ch_coeff1*rx_symbol2.conjugate())
                det_symbol1 = 2*(estimate1.real >= 0) - 1
                det_symbol2 = 2*(estimate2.real >= 0) - 1
                no_errors += 1*(tx_symbol1 != det_symbol1)+1*(tx_symbol2 != det_symbol2)
        ber[n] = 1.0*no_errors/(2*N)
        print "EbNodB:", EbNodB
        print "Numbder of errors:", no_errors
        print "Error probability:", ber[n]
plt.plot(EbNodB_range, ber, 'bo-')
plt.axis([0, 10, 0.001, 1])
plt.xscale('linear')
plt.yscale('log')
plt.xlabel('EbNo(dB)')
plt.ylabel('BER')
plt.grid(True)
plt.title('BPSK Modulation - Alamouti Scheme')
plt.show()
print time() - t, "seconds"
###################################################################################
```

9.5 Rayleigh Fading Envelope Generation

When wireless signals travel from a transmitter to a receiver they do so after reflection, refraction, diffraction and scattering from the environment. Very rarely is there a direct line of sight (LOS) between the transmitter and receiver. Thus, multiple time-delayed copies of the signal reach the receiver that combine constructively and destructively. In a sense the channel acts as an finite impulse response (FIR) filter. Furthermore, since the transmitter or receiver may be in motion the amplitude and phase of these replicas varies with time.

There are several methods to model the amplitude and phase of each of these components. We look at one method called the "Smiths Fading Simulator" which is based on Clark and Gans model. The simulator can be constructed using the following steps.

1. Define N the number of Gaussian RVs to be generated, f_m the Doppler frequency in Hz, f_s the sampling frequency in Hz, d_f the frequency spacing in Hz which is calculated as $d_f = 2f_m/(N-1)$ and M total number of samples in frequency domain which is calculated as $M = f_s/d_f$.
2. Generate two sequences of $N/2$ complex Gaussian random variables. These correspond to the frequency bins up to f_m. Take the complex conjugate of these sequences to generate the $N/2$ complex Gaussian random variables for the negative frequency bins up to $-f_m$.
3. Multiply the above complex Gaussian sequences g_1 and g_2 with square root of the Doppler Spectrum S generated from $-f_m$ to f_m. Calculate the spectrum at $-f_m$ and f_m by using linear extrapolation.
4. Extend the above generated spectra from $-f_s/2$ to $f_s/2$ by stuffing zeros from $-f_s/2$ to $-f_m$ and f_m to $f_s/2$. Take the IFFT of the resulting spectra X and Y resulting in time domain signals x and y.
5. Add the absolute values of the resulting signals x and y in quadrature. Take the absolute value of this complex signal. This is the desired Rayleigh distributed envelope with the required temporal correlation.

The MATLAB code for generating Rayleigh random sequence with a Doppler frequency of f_m Hz is given below.

```
#####################################################################################
#                    RAYLEIGH FADING ENVELOPE GENERATION
#                    N is the number of Gaussian RVs to be generated
#       fm is the Doppler frequency in Hz, fs is the sampling frequency in Hz
#          df is the frequency spacing which is calculated as df=(2*fm)/(N-1)
#  M is the total number of samples in frequency domain which is calculated as M=(fs/df)
#                           Copyright 2020 RAYmaps
#####################################################################################
from numpy import sqrt
from numpy.random import randn
import numpy as np
import matplotlib.pyplot as plt

N=20.0;
fm=70.0;
df=(2*fm)/(N-1);
fs=1000;
M=round(fs/df);
T=1.0/df;
Ts=1.0/fs;

# Generating first Gaussian RV set
g=randn(int(N/2))+1j*randn(int(N/2))
gc=np.conj(g)
gcr=gc[::-1]
g1=np.concatenate((gcr,g),axis=0)


# Generating second Gaussian RV set
g=randn(int(N/2))+1j*randn(int(N/2))
gc=np.conj(g)
gcr=gc[::-1]
g2=np.concatenate((gcr,g),axis=0)

# Generating the Doppler Spectrum
f=np.arange(-fm, fm+df, df)
S=1.5/(np.pi*fm*sqrt(1-(f/fm)**2))
S[0]=2*S[1]-S[2]
S[-1]=2*S[-2]-S[-3]

# Shaping the RV sequence g1 and taking IFFT
X=g1*sqrt(S);
X=np.concatenate((np.zeros(int((M-N)/2)), X), axis=0)
X=np.concatenate((X, np.zeros(int((M-N)/2))), axis=0)
x=np.abs(np.fft.ifft(X))
# Shaping the RV sequence g2 and taking IFFT
Y=g2*sqrt(S)
Y=np.concatenate((np.zeros(int((M-N)/2)), Y), axis=0)
Y=np.concatenate((Y, np.zeros(int((M-N)/2))), axis=0)
y=np.abs(np.fft.ifft(Y))

# Generating complex envelope
z=x+1j*y
r=np.abs(z)
```

```
# Plotting the envelope in the time domain
t=np.arange(0, T, Ts)
plt.plot(t, 10*np.log10(r/np.max(r)),'b')
plt.show()
plt.xlabel('Time(msecs)')
plt.ylabel('Envelope(dB)')
plt.grid(True)
plt.title('Rayleigh Fading')
###############################################################################
```

9.6 BER for BPSK-OFDM in Frequency Selective Channel

As the data rates supported by wireless networks continue to rise the bandwidth requirements also continue to increase (although spectral efficiency has also improved). Remember GSM technology which supported 125 channels of 200 kHz each, which was further divided among eight users using TDMA. Move on to LTE where the channel bandwidth could be as high as 20 MHz (1.4, 3, 5, 10, 15 and 20 MHz are standardized).

This advancement poses a unique challenge referred to as frequency selective fading. This means that different parts of the signal spectrum would see a different channel (different amplitude and different phase offset). Look at this in the time domain where the larger bandwidth means shorter symbol period causing intersymbol interference (as time-delayed copies of the signal overlap on arrival at the receiver).

The solution to this problem is OFDM that divides the wideband signal into smaller components each having a bandwidth of a few KHz. Each of these components experiences a flat channel. To make the task of equalization simple a cyclic prefix (CP) is added in the time domain to make the effect of fading channel appear as circular convolution. Thus simplifying the frequency domain equalization to a simple division operation.

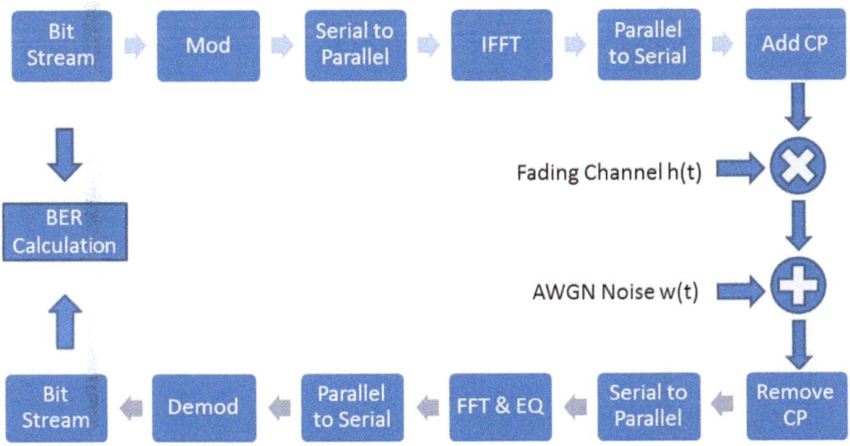

Fig. 9.2 OFDM Tx-Rx block diagram

Shown below is the Python code that calculates the bit error rate (BER) of BPSK-OFDM which is the same as simple BPSK in a Rayleigh flat fading channel. However, there is a caveat. We have inserted a CP which means we are transmitting more energy than simple BPSK. To be exact we are transmitting 1.25 (160/128) times more energy. This means that if this excess energy is accounted for, the performance of BPSK-OFDM would be 1 dB (10 * log10(1.25)) worse than simple BPSK in Rayleigh flat fading channel.

```
#####################################################################################
#       SIMULATION OF OFDM IN A TIME VARYING FREQUENCY SELECTIVE FADING CHANNEL
#                           n_cyc is the length of the cyclic prefix
#                           n_fft is the length of the FFT and IFFT
#                           n_sym is the number of OFDM symbols
#                    n_bits is the number of bits before the modulation operation
#                           n_tap is the length of the channel filter
#                                Eb is the energy per bit
#                         EbNo is the energy per bit to noise PSD ratio
#                               Copyright 2020 RAYmaps
#####################################################################################
from numpy import sqrt
from numpy.random import rand, randn
import numpy as np

n_cyc=32
n_fft=128
n_sym=10000
n_bits=n_fft*n_sym
n_tap=10
Eb=1.0
EbNodB=34.0
EbNo=10**(EbNodB/10.0)

s_in=2*(rand(n_bits)>=0.5)-1
s_est=np.zeros(n_bits)

for n in range (0, n_sym):
        s_ofdm=sqrt(n_fft)*np.fft.ifft(s_in[n*n_fft:(n+1)*n_fft])
        s_ofdm_cyc=np.concatenate((s_ofdm[n_fft-n_cyc:n_fft], s_ofdm), axis=0)
        ht=(1/sqrt(2))*(1/sqrt(n_tap))*(randn(n_tap)+1j*randn(n_tap));
        Hf=np.fft.fft(ht,n_fft)
        r_ofdm=np.convolve(s_ofdm_cyc,ht)
        r_ofdm_cyc=(r_ofdm[0:n_fft+n_cyc])
        wn=randn(n_fft+n_cyc)+1j*randn(n_fft+n_cyc)
        r_ofdm_cyc=r_ofdm_cyc+sqrt(Eb/(2*EbNo))*wn
        r_ofdm_cropped=r_ofdm_cyc[n_cyc:n_fft+n_cyc]
        s_est[n*n_fft:(n+1)*n_fft]=(1/sqrt(n_fft))*np.real((np.fft.fft(r_ofdm_cropped))/Hf)
s_out=2*(s_est>=0)-1
errors=(s_in!=s_out).sum()
ber=1.0*errors/n_bits

print('Total number of OFDM symbols', n_sym)
print('Total number of bits', n_bits)
print('Total number of bits in error', errors)
print('Energy per bit to noise PSD(dB)', EbNodB)
print('Bit Error Rate', ber)
#####################################################################################
```

Note:

1. Although we have shown the channel as a multiplicative effect in the figure above, this is only true for a single tap channel. For a multitap channel (such as the one used in the code above) the effect of the channel is that of a filter which performs convolution operation on the transmitted signal.
2. We have used a baseband model in our simulation and the accompanying figure. In reality the transmitted signal is up-converted before transmission by the antennas.
3. The above model can be easily modified for any modulation scheme such as QPSK or 16-QAM. The main difference would be that the signal would have a both a real part and an imaginary part, much of the simulation would remain the same. This would be the subject of a future article. For a MATLAB implementation of 64-QAM OFDM see Sect. 4.3.
4. Serial to parallel and parallel to serial conversion shown in the above figure was not required as the simulation was done symbol by symbol (one OFDM symbol in the time domain represented 128 BPSK symbols in the frequency domain).
5. The channel model in the above simulation is quasi-static, i.e., it remains constant for one OFDM symbol but then rapidly changes for the next, without any memory.

Questions and Numerical Problems

1. Compare the execution speed of Python with MATLAB and Octave for (a) Random number generation (b) For Loop implementation (c) Matrix manipulation.
2. Compare the execution speed of Python with MATLAB and Octave for bit error rate simulation of (a) BPSK (b) QPSK (c) 16-QAM.
3. Compare the execution speed of Python with MATLAB and Octave for bit error rate simulation of OFDM in AWGN.
4. Compare the execution speed of Python with MATLAB and Octave for bit error rate simulation of OFDM in frequency selective Rayleigh fading channel.
5. Compare the execution speed of Python with MATLAB and Octave for bit error rate simulation of Alamouti code in quasi-static, flat, Rayleigh fading channel.
6. Using the Python code for Rayleigh fading envelope generation, calculate the level crossing rate (LCR) and average fade duration (AFD).
7. Repeat the above for three cases of Doppler shift (a) Low (b) Medium (c) High. Comment on the dependency of LCR and AFD on Doppler frequency.
8. Plot the 3D radiation pattern of a dipole antenna using Python.
9. Plot the 3D radiation pattern of a Uniform Linear Array (ULA) using Python.
10. Plot the 3D radiation pattern of a (a) Rectangular array (b) Circular array, using Python.

Useful Links

1. BPSK Bit Error Rate Calculation Using Python
   https://www.raymaps.com/index.php/bpsk-bit-error-rate-calculation-using-python/
2. MATLAB vs Python Computational Speed
   https://www.raymaps.com/index.php/matlab-vs-python-computational-speed/
3. Alamouti—Transmit Diversity Scheme
   https://www.raymaps.com/index.php/alamouti-transmit-diversity-scheme-implemented-in-python/
4. Rayleigh Fading Envelope Generation
   https://www.raymaps.com/index.php/rayleigh-fading-envelope-generation-python/
5. BER for BPSK-OFDM in Frequency Selective Channel
   https://www.raymaps.com/index.php/ber-for-bpsk-ofdm-in-frequency-selective-channel/

MIX
Papier aus verantwortungsvollen Quellen
Paper from responsible sources
FSC® C105338

If you have any concerns about our products,
you can contact us on
ProductSafety@springernature.com

In case Publisher is established outside the EU,
the EU authorized representative is:
Springer Nature Customer Service Center GmbH
Europaplatz 3, 69115 Heidelberg, Germany

Printed by Libri Plureos GmbH
in Hamburg, Germany